Collins

INTERNATIONAL PRIMARY MATHS

T0173430

Student's Book 6

William Collins' dream of knowledge for all began with the publication of his first book in 1819. A self-educated mill worker, he not only enriched millions of lives, but also founded a flourishing publishing house. Today, staying true to this spirit, Collins books are packed with inspiration, innovation and practical expertise. They place you at the centre of a world of possibility and give you exactly what you need to explore it.

Collins. Freedom to teach.

Published by Collins
An imprint of HarperCollins*Publishers*
The News Building
1 London Bridge Street
London
SE1 9GF

HarperCollins *Publishers*
Macken House,
39/40 Mayor Street Upper,
Dublin 1
D01 C9W8 Ireland

Browse the complete Collins catalogue at
www.collins.co.uk

ISBN 978-0-00-836944-6

British Library Cataloguing-in-Publication Data
A catalogue record for this publication is available from the British Library.

Author: Paul Hodge
Series editor: Peter Clarke
Publisher: Elaine Higgleton
Product developer: Holly Woolnough
Project manager: Mike Harman (Life Lines Editorial Services)
Development editor: Tanya Solomons
Copyeditor: Catherine Dakin
Proofreader: Catherine Dakin
Cover designer: Gordon MacGilp
Cover illustrator: Ann Paganuzzi
Typesetter: Ken Vail Graphic Design
Illustrators: Ann Paganuzzi and QBS Learning
Production controller: Lyndsey Rogers
Printed and bound by Grafica Veneta S. P. A.

With thanks to the following teachers and schools for reviewing materials in development: Antara Banerjee, Calcutta International School; Hawar International School; Melissa Brobst, International School of Budapest; Rafaella Alexandrou, Pascal Primary Lefkosia; Maria Biglikoudi, Georgia Keravnou, Sotiria Leonidou and Niki Tzorzis, Pascal Primary School Lemessos; Taman Rama Intercultural School, Bali.

The publishers gratefully acknowledge the permission granted to reproduce the copyright material in this book. Every effort has been made to trace copyright holders and to obtain their permission for the use of copyright material. The publishers will gladly receive any information enabling them to rectify any error or omission at the first opportunity.
Cambridge International copyright material in this publication is reproduced under licence and remains the intellectual property of Cambridge Assessment International Education.

Contents

Number

Geometry and Measure

Statistics and Probability

How to use this book

This book is used towards the start of a lesson when your teacher is explaining the mathematical ideas to the class.

Key words
- The **key words** to use during the lesson are given. It's important that you understand the meaning of each of these words.

- An **objective** explains what you should know, or be able to do, by the end of the lesson.

Let's learn

This section of the Student's Book page **teaches** you the main mathematical ideas of the lesson. It might include pictures or diagrams to help you **learn**.

 An activity that involves thinking and working mathematically.

 An activity or question to discuss and complete in pairs.

Guided practice

Guided practice helps you to answer the questions in the Workbook. Your teacher will talk you through this question so that you can work independently with confidence on the Workbook pages.

HINT

Use the page in the Student's Book to help you answer the questions on the Workbook pages.

1 **Thinking and Working Mathematically** (TWM) involves thinking about the mathematics you are doing to gain a deeper understanding of the idea, and to make connections with other ideas. The TWM star at the back of this book describes the 8 ways of working that make up TWM. It also gives you some sentence stems to help you to talk with others, challenge ideas and explain your reasoning.

At the back of the book

Number

Lesson 1: **Counting on and back in fractions and decimals**

Key words
- counting on
- counting back
- ones boundary
- tenths boundary
- hundredths boundary

- Count on and back in decimal steps and fractions

Let's learn

When counting in decimals, what happens to the digits when the count crosses ones, tenths and hundredths boundaries?

Count on in steps of 0·3 from 2·3 to 3·8

2·3, 2·6, (2·9) 3·2, 3·5, 3·8

Tenths digit resets to zero and the count continues: 2·9, 3·0, 3·1, 3·2
+0·3

Ones digit increases by 1

Count back in steps of 0·04 from 4·23 to 4·03

4·23, 4·19, 4·15, (4·11) 4·07, 4·03

Hundredths digit resets to zero and the count continues: 4·11, 4·10, 4·09, 4·08, 4·07
−0·04

Tenths digit decreases by 1

Counting in fractions using a number line:

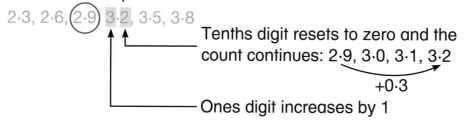

forward in $\frac{1}{3}$s

$1\frac{1}{3}$ $1\frac{2}{3}$ 2 $2\frac{1}{3}$ $2\frac{2}{3}$ 3 $3\frac{1}{3}$ $3\frac{2}{3}$ 4

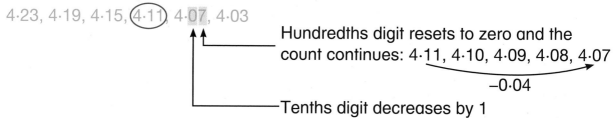

back in $\frac{1}{4}$s

$\frac{2}{4}$ $\frac{3}{4}$ 1 $1\frac{1}{4}$ $1\frac{2}{4}$ $1\frac{3}{4}$ 2 $2\frac{1}{4}$ $2\frac{2}{4}$

forward in $\frac{2}{5}$s

$3\frac{1}{5}$ $3\frac{2}{5}$ $3\frac{3}{5}$ $3\frac{4}{5}$ 4 $4\frac{1}{5}$ $4\frac{2}{5}$ $4\frac{3}{5}$ $4\frac{4}{5}$

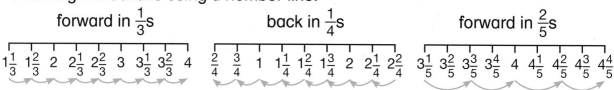

👥 Calculate the following:

$3·5 + 0·2 + 0·2 + 0·2 + 0·2 =$

$5·67 − 0·03 − 0·03 − 0·03 =$

$4·075 − 0·004 − 0·004 =$

$2\frac{1}{4} + \frac{1}{4} + \frac{1}{4} + \frac{1}{4} + \frac{1}{4} =$

$9\frac{2}{3} − \frac{1}{3} − \frac{1}{3} − \frac{1}{3} − \frac{1}{3} =$

$6\frac{1}{5} + \frac{3}{5} + \frac{3}{5} + \frac{3}{5} =$

Guided practice

A ribbon is 6·87 metres long. If four equal lengths of 0·04 metres are cut from the ribbon, what length remains?

$6·87 − 0·04 − 0·04 − 0·04 − 0·04$

6·87   6·83   6·79   6·75   6·71

−0·04 −0·04 −0·04 −0·04

The answer is: *6·71 metres*

Lesson 2: **Counting on and back beyond zero**

Key words
* counting on
* counting back
* negative number

* Count on and back in steps of whole numbers and fractions from different numbers including negative numbers

Let's learn

Knowing how to count on or back in whole numbers helps you to count in decimals and fractions.

Count back in 5s from 11

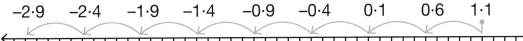

Count back in 0·5s from 1·1

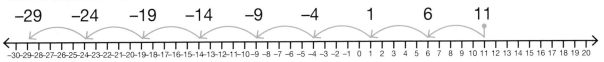

Count on in 6s from −25

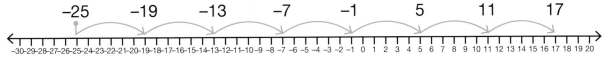

Count on in 0·6s from −2·5

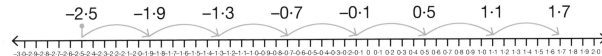

A bowl of water is placed in a freezer.

The temperature of the water drops by 0·4 degrees every minute. If the temperature reading at the start is 1·1°C, what will the reading be after these times: 2 minutes, 4 minutes, 7 minutes?

List the temperature after every minute up to 12 minutes. Generalise: is there a pattern? Will any of the readings have a tenths digit that is even? Predict whether a temperature of −5·4°C will be in this list. How do you know?

> ### Guided practice
> The balance of a bank account is $9·30.
> Charges on the account mean that the balance reduces by $0·80 each day.
> What will the balance be after 6 days?
> $9·3 − 0·8 − 0·8 − 0·8 − 0·8 − 0·8 − 0·8 = \$4·50$

Number

Lesson 3: Finding the position-to-term rule

* Know the rule of a sequence, and use the position of a term in the sequence to calculate its value

Let's learn

A sequence is a list of values in a certain order. Each number in a sequence is called a **term**.

Each term in a sequence has a **position**. The first term is in position 1, the second term is in position 2, and so on.

Position	1	2	3	4	5	6	7	8
Term	4	8	12	16	20	24	28	32

In the table above, you can see that the terms are multiples of 4. The term in position 3 is 12. The term in position 7 is 28.

The **position-to-term rule** tells you how to find each term.

For the sequence of multiples of 4, the position in the sequence multiplied by 4 gives the term.

The term in position 5 can be found with $5 \times 4 = 20$.

The term in position 8 can be found with $8 \times 4 = 32$.

a What is the value of the terms in the following positions: 2, 5 and 7?

Position	1	2	3	4	5	6	7	8
Term	9	18	27	36	45	54	63	72

b What is the position of the term with a value of:

   i 36?

   ii 9?

   iii 54?

c What is the position-to-term rule?

d How would you use the rule to find the 9th, 11th and 14th terms?

Guided practice

Position	1	2	3	4	5	6
Term	7	14	21	28	35	42

What is the value of the:

a 8th term?   **b** 11th term?   **c** 14th term?

From the table, I can see that the value of the terms follows the 7 times table.

I know the position-to-term rule: multiply by 7.

a Value of the 8th term: $8 \times 7 = 56$

b Value of the 11th term: $11 \times 7 = 77$

c Value of the 14th term: $14 \times 7 = 98$

Lesson 4: Finding terms of a square number sequence

Key words
* sequence
* term
* position-to-term rule
* square number

* Use the position of a term in the sequence of square numbers to calculate its value

Let's learn

The **square number** sequence

Position	1	2	3	4	5
Pattern	•	•• ••	••• ••• •••	•••• •••• •••• ••••	••••• ••••• ••••• ••••• •••••
Term	1 (1 × 1)	4 (2 × 2)	9 (3 × 3)	16 (4 × 4)	25 (5 × 5)

The **position-to-term rule** tells you how to find each term.

For the sequence of square numbers, the position in the sequence squared gives the term.

The term in position 6 can be found with $6^2 = 36$.

The term in position 8 can be found with $8^2 = 64$.

Seeds are planted in plots that have neat rows and columns.

For each plot, the number of seeds in each row is the same as the number of seeds in each column.

The number of seeds in the row of the first plot is 1. The number of seeds in the row of the second plot is 2. The number of seeds in the row of the third plot is 3, and so on.

How many seeds in total will be needed for the first six plots?

Guided practice

Position	6	7	8	9
Term	36	49	64	81

Given the terms and their positions, what can you say about this sequence? What are the 11th and 12th terms in the sequence?

I can see that the terms 36, 49, 64 and 81 form part of the sequence of square numbers.

I know the position-to-term rule: square the position number.

Value of 11th term: $11^2 = 11 \times 11 = 121$

Value of 12th term: $12^2 = 12 \times 12 = 144$

Lesson 1: Adding positive and negative numbers (1)

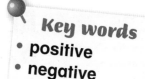

Key words
- positive
- negative

- Use a number line to add positive and negative numbers

Let's learn

Example: −19 + 5 = −14

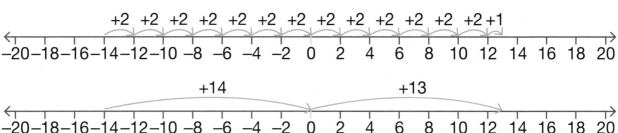

Example: −14 + 27 = 13

A flying fish travels up from the sea bed and flips into the air. Work out the heights reached above sea level by the fish.

Depth of sea bed	−16 m	−14 m	−18 m	−26 m	−31 m
Height travelled	18 m	19 m	23 m	38 m	44 m

Guided practice

Matsu's bank account is overdrawn at −$18. This means she has spent more money than she actually had in her account.

She pays in $22. How much money does Matsu have in her account now? Estimate first. Is she still overdrawn? How do you know?

Sara estimates that the amount she will have after paying in $22 will be positive as 22 > 18. −18 + 22 = 4

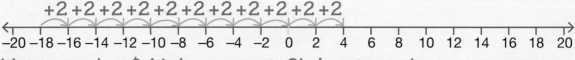

Matsu now has $4 in her account. She's not overdrawn any more as the amount is positive.

Lesson 2: **Adding positive and negative numbers (2)**

* Use a number scale to add positive and negative numbers

Let's learn

$-12 + 25 = 13$

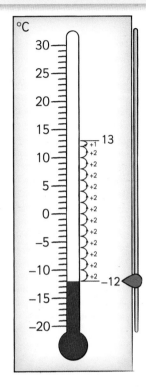

Work out the new temperatures after two rises.

Start temperature	−16 °C	−17°C	−18 °C	−19 °C	−14 °C
Increase	14 degrees	13 degrees	13 degrees	17 degrees	8 degrees
New temperature	°C	°C	°C	°C	°C
Increase	12 degrees	16 degrees	22 degrees	25 degrees	33 degrees
New temperature	°C	°C	°C	°C	°C

Guided practice

$-17 + 23 =$ Estimate first. Since 23 > 17, the answer will be positive.

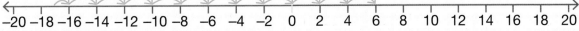

$+1 +2 +2 +2 +2 +2 +2 +2 +2 +2 +2 +2$

−20 −18 −16 −14 −12 −10 −8 −6 −4 −2 0 2 4 6 8 10 12 14 16 18 20

$-17 + 23 = 6$

Number

Lesson 3: **Identifying values for symbols in addition calculations**

- Find the value of unknown values in calculations that are represented by symbols

Let's learn

You can use symbols, such as letters of the alphabet, to represent unknown values in a calculation.

$23 + a = 32$ $\qquad$ $b + 16 = 39$ $\qquad$ $c + 48 = 62$

Use the inverse operation to find the unknown.

$a = 32 - 23 = 9$ $\qquad$ $b = 39 - 16 = 23$ $\qquad$ $c = 62 - 48 = 14$

Ekon visits a shop.

He buys a pack of colouring pens and a diary for $41.

The colouring pens cost $17.

Write a number sentence to represent the calculation. Use a letter to represent the unknown value.

What is the price of the diary? How did you work it out?

Guided practice

A suitcase and a carrier bag have a combined mass of 33 kg.

If the carrier bag has a mass of 6 kg, what is the mass of the suitcase?

$6 + s = 33$

$s = 33 - 6$

$s = 27$

The mass of the suitcase is 27 kg.

Lesson 4: **Identifying values for symbols in subtraction calculations**

Key words
- unknown number
- inverse operation

Number

- Find the value of unknown values in calculations that are represented by symbols

Let's learn

You can use symbols, such as letters of the alphabet, to represent unknown values in a calculation.

$a - 12 = 17$ $\qquad$ $43 = b - 15$ $\qquad$ $35 - c = 18$

Use the inverse operation to find the unknown.

$a = 17 + 12 = 29$ $\qquad$ $b = 43 + 15 = 58$ $\qquad$ $35 = 18 + c$
$35 - 18 = c = 17$

Amani visits a shop.

She buys an umbrella.

She pays with two $20 notes and receives $18 in change.

What was the price of the umbrella?

Write a number sentence to represent the calculation.
Use a letter to represent the unknown value.

How did you work it out?

Guided practice

Joe has prepared a jug of squash for a party.

Before the party begins, he pours out 180 ml of squash and drinks it.

565 ml of squash remains in the jug. How much was in the jug to begin with?

Let *s* represent the original amount of squash in the jug.

$565 = s - 180$

$565 + 180 = s$ (using the inverse operation of addition)

Estimate $565 + 180$ by rounding: $550 + 200 = 750$

$s = 745$

The original amount of squash in the jug was 745 ml.

Lesson 1: **Subtracting positive and negative integers**

Key word
• difference

• Find the difference between positive and negative numbers

Let's learn

What is the difference between 3 and –6?

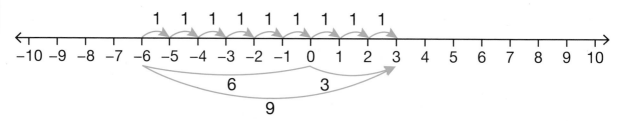

👥 The table compares the average temperature between Canada and the United Kingdom over five days.

Work out the difference between the two temperatures in degrees.

	Monday	Tuesday	Wednesday	Thursday	Friday
Canada (average temperature, °C)	–8	–9	–7	–11	–13
UK (average temperature, °C)	11	13	15	12	16

Guided practice

Feechi reads the temperature on a thermometer in a glass of frozen water. The temperature is –8°C.

He leaves the water by a radiator and takes the temperature a few hours later. It is 9°C.

What is the difference between the two temperature readings? Estimate first.

Estimate 8 + 9 first by rounding: 10 + 10 = 20.

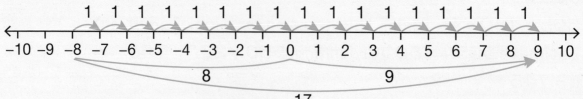

The difference between –8 and 9 is 17.

Lesson 2: **Subtracting two negative integers**

- Find the difference between two negative numbers

Number

Let's learn

The difference between –12 °C and –5 °C is 7 degrees.

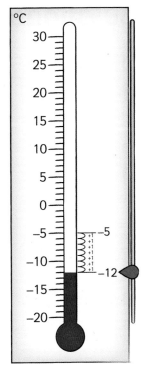

Five people go shopping. They each spend a small amount of money.
Look at their bank accounts and work out how much each person spent.
Who has spent the most?

	Lana's account	Logan's account	Tonya's account	Atique's account	Maisie's account
Morning account balance	–$2	–$4	–$1	–$3	–$5
Evening account balance	–$11	–$14	–$13	–$14	–$13

Guided practice

Use the number line to work out the difference between –3 and –14.

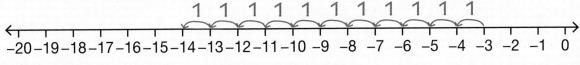

The difference between –3 and –14 is 11.

Number

Lesson 3: **Identifying values of variables in calculations (1)**

- Identify the value of variables in calculations

Let's learn

$a + b = 10$

Solutions including zero:

$a = 0, b = 10$ (as $0 + 10 = 10$) $a = 1, b = 9$ (as $1 + 9 = 10$)

$a = 2, b = 8$ (as $2 + 8 = 10$) $a = 3, b = 7$ (as $3 + 7 = 10$)

$a = 4, b = 6$ (as $4 + 6 = 10$) $a = 5, b = 5$ (as $5 + 5 = 10$)

$a = 6, b = 4$ (as $6 + 4 = 10$) $a = 7, b = 3$ (as $7 + 3 = 10$)

$a = 8, b = 2$ (as $8 + 2 = 10$) $a = 9, b = 1$ (as $9 + 1 = 10$)

$a = 10, b = 0$ (as $10 + 0 = 10$)

Don't forget there will also be some negative number solutions. Here are some examples:

$-1 + 11 = 10$ $-2 + 12 = 10$ $-3 + 13 = 10$ $-4 + 14 = 10$

A carpenter saws a length of timber 120 cm long into two pieces, x and y.

The lengths of the pieces can be only be in multiples of 10 cm.

Conjecture: what are all the possible values for x and y?

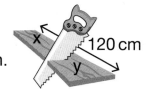

Guided practice

6 kg of sand is split between two buckets labelled C and D.

Each bucket can have any whole number of kilograms of sand.

What different combinations are possible? How many are there?

I can write a number sentence for this: $c + d = 6$ kg

Where c is the amount of sand in bucket C, and d is the amount of sand in bucket D.

Next, I substitute different values for c and d.

$c = 1$ kg, $d = 5$ kg (as $1 + 5 = 6$) $c = 2$ kg, $d = 4$ kg (as $2 + 4 = 6$)

$c = 3$ kg, $d = 3$ kg (as $3 + 3 = 6$) $c = 4$ kg, $d = 2$ kg (as $4 + 2 = 6$)

$c = 5$ kg, $d = 1$ kg (as $5 + 1 = 6$)

There are five different combinations.

Lesson 4: **Identifying values of variables in calculations (2)**

Key words
- variable
- equation
- formula (formulae)

Number

- Use a simple formula for given values

Let's learn

The number of green and purple counters in a bag is given by the formula $T = g + p$. T is the total number of counters, g is the number of green counters and p is the number of purple counters.

Number of green counters (g)	6	14	18	26	47
Number of purple counters (p)	8	13	24	36	37
Total (T)	6 + 8 = 14	14 + 13 = 27	18 + 24 = 42	26 + 36 = 62	47 + 37 = 84

Write the formula for finding the perimeter of a square.

Copy the table. Use the formula to fill in the missing numbers.

Sides (cm)	2	5	8	13	18	34
Perimeter						

Guided practice

Tai works in a café. He is paid $78 per day, plus whatever he receives in tips.

The amount he earns each day is given by the formula $E = \$78 + T$ where E is amount earned and T is tips.

Use the formula to complete the table.

Remember to estimate first, for example, 78 + 44 will be approximately 80 + 45 = 125.

T ($)	5	9	13	26	33	44
E ($)	78 + 5 = 83	78 + 9 = 87	78 + 13 = 91	78 + 26 = 104	78 + 33 = 111	78 + 44 = 122

Number

Lesson 1: **Common multiples**

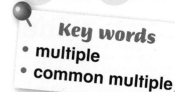

Key words
• multiple
• common multiple

• Understand and find common multiples

Let's learn

To find **common multiples** of two numbers, list the multiples of each number and find the numbers that appear in both lists.

Example:

Multiples of 4: 4, 8, 12, 16, 20, 24, 28, 32, 36, …

Multiples of 6: 6, 12, 18, 24, 30, 36, 42, …

The first three common multiples of 4 and 6 are 12, 24 and 36.

Nathan has falafels for breakfast every four days and meatballs for lunch every five days.

3

Conjecture: how often will he have falafels and meatballs

4 on the same day?

Number

Lesson 2: **Common factors**

• Understand and find common factors

Key words
• factor
• common factor

Let's learn

To find **common factors** of two numbers, list the factors of each number and find the numbers that both lists have in common.

Example:

Factors of 12: ①② 3,④ 6, 12

Factors of 16: ①②④ 8, 16

The common factors of 12 and 16 are 1, 2 and 4.

Hassam has 36 football stickers and Lucy has 24 netball stickers.

They place the stickers into piles so that each pile has exactly the same number of football stickers, and each pile has exactly the same number of netball stickers.

What is the largest number of piles that can be made?

Guided practice

What are the common factors of 40 and 64?

Factors of 40: ① ② ④ 5, ⑧ 10, 20, 40

Factors of 64: ① ② ④ ⑧ 16, 32, 64

The common factors of 40 and 64 are 1, 2, 4 and 8.

19

Lesson 3: **Tests of divisibility by 3, 6 and 9**

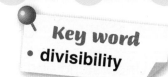

Key word
• divisibility

• Test numbers for divisibility by 3, 6 and 9

Let's learn

Is 732 divisible by 3, 6 and 9?

Divisible by …	Rule	Check	Yes/No
9	The sum of the digits is a multiple of 9.	$7 + 3 + 2 = 12$	No
3	The sum of the digits is a multiple of 3.	See above	Yes
6	The number is divisible by 2 and 3.	3: Confirmed 2: Even number	Yes

Find five 3-digit numbers that are divisible by 3, 6 and 9.

8 As you test each number, decide which divisibility test it is best to apply first, then which test it is best to apply second. Why are you convinced by your order?

Guided practice

Which of these numbers are divisible by 6?

347   565   984   581   348

Number	Divisible by 2? (Is the number even?)	Divisible by 3 (Do the digits add up to a multiple of 3?)	Yes/No
347	No	No need to test	No
565	No	No need to test	No
984	Yes	$9 + 8 + 4 = 21$	Yes
581	No	No need to test	No
348	Yes	$3 + 4 + 8 = 15$	Yes

984 and 348 are divisible by 6.

Lesson 4: **Cube numbers**

- Use square numbers to recognise cube numbers

Key words
- square number
- cube number

Number

Let's learn

You can use a square number to find a **cube number**.

Number	Square number	Cube number
3	9 (3 × 3)	9 × 3 = 27
4	16 (4 × 4)	16 × 4 = 64
5	25 (5 × 5)	25 × 5 = 125

Example: Jin is asked to find the cube of 2.

He knows that 4 is the square of 2.

To get the cube of 4, she multiples by 2 again: 4 × 2 = 8

The cube of 2 is 8.

Find a square number that is also a cube number.

Guided practice

What is the value of 5^3?

$5^3 = 5 \times 5 \times 5 = 25 \times 5 = 125$

Number

Lesson 1: **Simplifying calculations (1)**

* Simplify calculations using the properties of number

Let's learn

You can use the properties of number to simplify calculations.

Commutative property

You can change the order of the augend and the addend.

You can change the order of the multiplier and multiplicand.

Example: $7 + 5 = 5 + 7$

Example: $4 \times 6 = 6 \times 4$

Associative property

You can change the grouping of the augend and addends.

Example: $8 + 14 + 22 = 8 + 22 + 14$

You can change the order of the terms in a multiplication.

Example: $32 \times 5 = 4 \times 8 \times 5$

$$4 \times 40 = 160$$

Or apply both associative and commutative laws.

Example: $32 \times 5 = 4 \times 8 \times 5 = 4 \times 5 \times 8$

$$20 \times 8 = 160$$

Distributive property

You can use the distributive property to break apart a multiplication.

Example: $8 \times 64 = 8 \times 60 + 8 \times 4$

👥 Simplify and solve the following calculations.

    a $15 \times 7 \times 4 \times 3$ **b** 8×73 **c** $36 + 27 + 44 + 63$

Guided practice

Simplify and solve the calculation. Describe the property of number you use.
$4 \times 6 \times 5 \times 3$

$4 \times 6 \times 5 \times 3 = 4 \times 5 \times 6 \times 3 = 20 \times 18$

$$= 10 \times 2 \times 18 = 10 \times 36 = 360$$

I used the associative property.

Lesson 2: **Simplifying calculations (2)**

Key words
* **commutative property**
* **associative property**
* **distributive property**
* **multiplier**
* **multiplicand**
* **augend**
* **addend**
* **term**

* Simplify calculations using the properties of number

Let's learn

You can use the properties of number to simplify calculations when solving real world problems.

Example: Crates are stacked in a warehouse in 3 layers of 25.

There are 4 similar stacks of crates, plus a loose set of 5 crates.

How many crates are there in total?

Calculation to solve:

$25 \times 3 \times 4 + 5$

$= 25 \times 4 \times 3 + 5$ (reordering terms of multiplication using the associative property)

$= 100 \times 3 + 5$ (solving 25×4)

$= 300 + 5$ (solving 100×3)

$= 305$

There are 305 crates in total.

Nadia works at the supermarket. She takes several payments in a period of 10 minutes.

The receipts she gives to the customers are shown here.

What is the total payment she receives?

Receipt	Receipt	Receipt
$43	$58	$125
Receipt	Receipt	Receipt
$75	$27	$32

Guided practice

Write a calculation for the problem. Simplify and solve it.

There are 12 beetles on each leaf of a tree.

There are 5 branches on the tree, each with 15 leaves.

There are 6 identical trees. How many beetles are there in total?

$6 \times 5 \times 15 \times 12 = 6 \times 15 \times 5 \times 12$ (associative property)

$= 90 \times 60$

$= 5400$

There are 5400 beetles.

Lesson 3: **Using brackets (1)**

- Understand that when completing a calculation that includes brackets, operations in brackets must be completed first

Key words
- brackets
- order of operations

Let's learn

Order of Operations

B ( ) Brackets
M × Multiplication
D ÷ Division
A + Addition
S − Subtraction

Example: $4 \times (8 + 13)$

$= 4 \times 21$ (brackets before multiplication)

$= 84$

Example: $25 \div (11 - 6)$

$= 25 \div 5$ (brackets before division)

$= 5$

👥 Simplify and solve the following calculations.

    a $8 \times (6 + 4)$ **b** $64 \div (16 - 8)$

Guided practice

Simplify and solve the calculation.

$15 \times (6 + 3)$

$15 \times (6 + 3) = 15 \times 9$

$= 135$

Number

Lesson 4: **Using brackets (2)**

• Understand that when completing a calculation that includes brackets, operations in brackets must be completed first

Let's learn

Elisa is asked to solve the following problem.

  Baskets of fruit contain 8 apples and 2 oranges.

  How many pieces of fruit are there in 12 baskets?

Elisa writes the calculation: $12 \times 8 + 2$.

Following the order of operations,
Elisa completes the multiplication first: $12 \times 8 + 2$

$$= 96 + 2 = 98$$

This is incorrect as the calculation only included one group of 2 apples.

As there are 12 lots of 8 apples **and** 12 lots of 2 oranges,
the correct calculation is:

$$12 \times (8 + 2)$$
$$= 12 \times 10$$
$$= 120$$

Parents have been invited to a school picnic.

Each person receives 5 white bread sandwiches and 3 brown bread sandwiches.

296 sandwiches were prepared in total. How many people attended the picnic?

Write and solve a calculation that represents the problem.

Guided practice

Write a calculation for the problem. Simplify and solve it. Use brackets to show which part of the calculation needs to be solved first.

18 construction bricks are placed in each of 8 buckets.

7 bricks are removed from each bucket. How many bricks are there in total in the buckets?

$8 \times (18 - 7) = 8 \times 11 = 88$

There are 88 bricks in total in the buckets.

Lesson 1: **Multiplying by 1-digit numbers (1)**

Key word
• expanded written method

Number

• Use the expanded written method to multiply numbers up to 10 000 by 1-digit whole numbers

Let's learn

$5367 \times 6 =$

Expanded written method

Estimate by rounding: $5000 \times 6 = 30\,000$

```
      5 3 6 7
  ×           6
          4 2   (7 × 6)
        3 6 0   (60 × 6)
      1 8 0 0   (300 × 6)
  + 3 0 0 0 0   (5000 × 6)
    3 2 2 0 2
        1   1
```

An aircraft makes 7 journeys, each with a distance of 3484 kilometres.

How many kilometres does the aircraft fly in total?

Guided practice

$2783 \times 4 =$

Estimate: $3000 \times 4 = 12\,000$

```
      2 7 8 3
  ×           4
          1 2   (3 × 4)
        3 2 0   (80 × 4)
      2 8 0 0   (700 × 4)
  +   8 0 0 0   (2000 × 4)
    1 1 1 3 2
          1
```

Lesson 2: **Multiplying by 1-digit numbers (2)**

• Use the formal written method to multiply numbers up to 10 000 by 1-digit whole numbers

Let's learn

$2741 \times 6 =$

Formal written method (short multiplication)

Estimate by rounding: $3000 \times 6 = 18\,000$

```
    2 7 4 1
×         6
  1 6 4 4 6
    4 2
```

A farmer plants 7657 seeds in each of 6 fields.

How many seeds does she plant in total?

Use the formal written method to calculate the answer.

Guided practice

$488 \times 7 =$

Estimate: $500 \times 7 = 3500$

```
    4 8 8
×         7
  3 4 1 6
    6 5
```

Number

Lesson 3: **Multiplying by 2-digit numbers (1)**

Key words
• partitioning
• grid method
• expanded written method

• Use partitioning and the grid method to multiply numbers up to 10 000 by 2-digit numbers

Let's learn

$2741 \times 6 =$

a Multiply numbers by 2-digit multiples of 10 in two steps:

Step 1   first multiply by the tens digit of the 2-digit number

Step 2   multiply the product of Step 1 by 10.

Example: What is 4374×30?

Estimate by rounding: $4000 \times 30 = 120\,000$

$4374 \times 3 = (4000 \times 3) + (300 \times 3) + (70 \times 3) + (4 \times 3)$

$\qquad = 12\,000 + 900 + 210 + 12$

$\qquad = 13\,122$

$4374 \times 30 = 4374 \times 3 \times 10$

$\qquad = 13\,122 \times 10$

$\qquad = 131\,220$

b Multiply numbers by 2-digit numbers using the grid method

Example: $689 \times 43 =$

Estimate by rounding: $700 \times 40 = 28\,000$

×	600	80	9	
40	24 000	3200	360	27 560
3	1800	240	27	+ 2067
			Answer:	29 627

6387 litres of water flows over a waterfall every minute.

How many litres flow in half an hour?

How many litres flow in 47 minutes?

Guided practice

$3739 \times 60 =$

$3739 \times 60 = 3739 \times 6 \times 10.$

$3739 \times 6 = (3000 \times 6) + (700 \times 6) + (30 \times 6) + (9 \times 6)$

$\qquad = 18\,000 + 4200 + 180 + 54$

$\qquad = 22\,434$

$3739 \times 6 \times 10 = 22\,434 \times 10$

$\qquad = 224\,340$

Lesson 4: **Multiplying by 2-digit numbers (2)**

Key words
- expanded written method
- formal written method (long multiplication)

- Use the formal written method of long multiplication to multiply numbers up to 10 000 by 2-digit numbers

Let's learn

5364 × 32 =

Expanded written method

Estimate by rounding:
5000 × 30 = 150 000

```
        5  3  6  4
×             3  2
     1  0  7  2  8   (5364 × 2)
  1  6  0  9  2  0   (5364 × 30)
  1  7  1  6  4  8
        1
```

Formal written method (long multiplication)

```
           5  3  6  4
×                3  2
        1  0  7¹ 2  8
     1  6¹ 0¹ 9¹ 2  0
     1  7  1  6  4  8
           1
```

67 runners are in a race.

They each run 3743 metres.

What is the combined distance run by all 67 runners?

Guided practice

728 × 37 =

Estimate: 700 × 40 = 28 000

```
        7  2  8
×          3  7
     5  0¹ 9⁵ 6
  2  1  8² 4  0
  2  6  9  3  6
        1
```

Number

Lesson 1: **Dividing 2-digit numbers by 1-digit numbers (1)**

- Use the 'chunking' method to divide 2-digit numbers by 1-digit numbers

Let's learn

You can model division by a 'chunking' method. Here, place value counters are used to represent a 2-digit dividend.

Example: 76 ÷ 3 = 25 r 1

Estimate by rounding: 75 ÷ 3 = 25

Begin by dividing in one 'chunk', finding how many groups of 10 times the divisor are in the dividend.

Next, regroup the 10s and 1s of the dividend and find the number of groups of the divisor.

Any remainder can be represented in the answer as a whole number or fraction of the divisor.

10s	1s
10 10 10 10	1 1 1 1
10 10 10	1 1

10s	1s
10 10 10	1 1 1 1
10 10 10	1 1 1 1
	1 1 1 1
	1 1 1 1

2 groups of 30 or 20 groups of 3

5 groups of 3 Remainder: 1

Logan has a piece of material that is 97 cm long. He cuts it into 6 strips of equal length.

What is the length of each strip? Give your answer as a mixed number.

Guided practice

99 ÷ 4 =

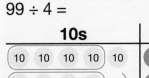

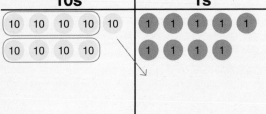

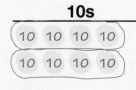

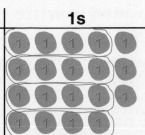

10s	1s
10 10 10 10 10	1 1 1 1 1
10 10 10 10	1 1 1 1

10s	1s
10 10 10 10	1 1 1 1 1
10 10 10 10	1 1 1 1 1
	1 1 1 1 1
	1 1 1 1

2 groups of 40 or 20 groups of 4

4 groups of 4 Remainder 3

$99 \div 4 = 24 \text{ r } 3 \text{ or } 24\frac{3}{4}$

Number

Lesson 2: Dividing 2-digit numbers by 1-digit numbers (2)

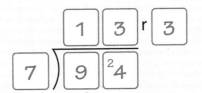
Key words
• dividend
• divisor
• expanded written method
• short division

• Use the expanded written method and short division to divide 2-digit numbers by 1-digit numbers

Let's learn

$99 \div 4 =$

Estimate by rounding: $100 \div 4 = 25$

Expanded written method

```
     2  4  r  3
  4 ) 9  9
  -  8  0     (4 × 20)
     1  9
  -  1  6     (4 × 4)
        3
```

Formal written method (short division)

```
     2   4  r  3
  4 ) 9  ¹9
```

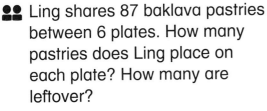 Ling shares 87 baklava pastries between 6 plates. How many pastries does Ling place on each plate? How many are leftover?

Calculate, using short division.

Guided practice

$94 \div 7 =$

Estimate: Between 10 and 20

$94 \div 7 = 13 \text{ r } 3 \text{ or } 13\frac{3}{7}$

```
        1   3  r  3
  7 ) 9  ²4
```

31

Number

Lesson 3: Dividing 3-digit numbers by 1-digit numbers (1)

Key words
- dividend
- divisor
- place value
- chunking method

- Use the 'chunking' method to divide 3-digit numbers by 1-digit numbers

Let's learn

$327 \div 5 = 65$ r 2

Estimate by rounding: $300 \div 5 = 60$

Begin by dividing in one 'chunk', finding how many groups of 100 times the divisor are in the dividend. For this division, there are no groups of 500 (100×5).

Next, regroup the 100s and 10s of the dividend. Then find the number of groups of 50 (10 times the divisor) in the dividend.

Next, regroup the 10s and 1s of the dividend and find the number of groups of the divisor. 65 groups of 5, remainder 2 or $\frac{2}{5}$.

100s	10s	1s

100s	10s	1s

6 groups of 50 or 60 groups of 5	5 groups of 5 Remainder: 2

It takes Sam 8 days to build a wall of 536 bricks.

If Sam lays the same number of bricks each day, how many bricks does he lay in one day?

Guided practice

$245 \div 7 =$

100s	10s	1s	100s	10s	1s

3 groups of 70 or 30 groups of 7

5 groups of 7

$245 \div 7 = 35$

Lesson 4: **Dividing 3-digit numbers by 1-digit numbers (2)**

Number

• Use the expanded written method and short division to divide 3-digit numbers by 1-digit numbers

Let's learn

$438 \div 5 =$

Estimate by rounding: $400 \div 5 = 80$

Expanded written method

```
        0  8  7  r  3
   5 ) 4  3  8
     -  4  0  0    (5 × 80)
           3  8
     -     3  5    (5 × 7)
              3
```

Formal written method (short division)

```
        0   8   7   r   3
   5 ) 4  ⁴3  ³8
```

A long distance run of 236 km is divided into 7 equal stages. Each runner completes their stage before handing over to the next runner.

Calculate the length of each stage using short division. Write your answer as a mixed number.

$345 \div 4 =$

Estimate: Between 80 and 90

$345 \div 4 = 86 \text{ r } 1 \text{ or } 86\frac{1}{4}$

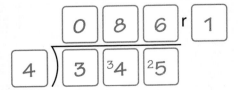

```
        0   8   6   r   1
   4 ) 3  ³4  ²5
```

Number

Lesson 1: Dividing 2-digit numbers by 2-digit numbers (1)

Key words
- dividend
- divisor
- quotient
- expanded written method

- Use the expanded written method to divide 2-digit numbers by 2-digit numbers

Let's learn

$81 \div 27 = 3$

Estimate by rounding: $75 \div 25 = 3$

Step 1

$$27 \overline{)8 \quad 1}$$

Step 2

$$27 \overline{)8 \quad 1}$$
$$- \quad 2 \quad 7 \quad (27 \times 1)$$
$$\overline{ 5 \quad 4}$$

Step 3

$$27 \overline{)8 \quad 1}$$
$$- \quad 2 \quad 7 \quad (27 \times 1)$$
$$\overline{ 5 \quad 4}$$
$$- \quad 2 \quad 7 \quad (27 \times 1)$$
$$\overline{ 2 \quad 7}$$

Step 4

$$\overset{\displaystyle 3}{27 \quad 8 \quad 1}$$
$$- \quad 2 \quad 7 \quad (27 \times 1)$$
$$\overline{ 5 \quad 4}$$
$$- \quad 2 \quad 7 \quad (27 \times 1)$$
$$\overline{ 2 \quad 7}$$
$$- \quad 2 \quad 7 \quad (27 \times 1)$$
$$\overline{ \qquad 0}$$

👥 Israa has \$85 to spend. She wants to buy scarves priced at \$17 for each of her friends.

How many friends can she buy scarves for?

Use the expanded written method of division to calculate the answer.

Guided practice

$78 \div 26 =$

Estimate: 3

There are 3 groups of 26 in 78.

$78 \div 26 = 3$

$$\overset{\boxed{3}}{\boxed{26}\overline{)\boxed{7}\ \boxed{8}}}$$
$$- \quad \boxed{2}\ \boxed{6} \qquad \boxed{26} \times \boxed{1}$$
$$\overline{ \boxed{5}\ \boxed{2}}$$
$$- \quad \boxed{2}\ \boxed{6} \qquad \boxed{26} \times \boxed{1}$$
$$\overline{ \boxed{2}\ \boxed{6}}$$
$$- \quad \boxed{2}\ \boxed{6} \qquad \boxed{26} \times \boxed{1}$$
$$\overline{ \boxed{0}}$$

Number

Lesson 2: **Dividing 2-digit numbers by 2-digit numbers (2)**

- Use the compact form of the expanded written method to divide 2-digit numbers by 2-digit numbers

Key words
- dividend
- divisor
- quotient
- compact form of the expanded written method
- trial and improvement method

Let's learn

$57 \div 19 =$

Estimate by rounding: $60 \div 20 = 3$

Estimation method 1: Trial and improvement

$19 \times 2 = 38$ (no), $19 \times 3 = 57$ (yes)

Estimation method 2: Focus on the ones digit

How many groups of the ones digit of the dividend give a number with the ones digit of the dividend?

$9 \times \mathbf{3} = 27$ (7 is the ones digit of the dividend)
Does $19 \times \mathbf{3} = 57$? (yes)

$$
\begin{array}{r}
3 \\
19 \overline{)5\ \ 7} \\
-\ \ 5\ \ 7 \quad (19 \times 3) \\
\hline
0 \\
\hline
57 \div 19 = 3
\end{array}
$$

👥 A paddling pool contains 84 litres of water.

Buckets have a capacity of 12 litres. How many bucketfuls will it take to remove all the water from the paddling pool?

Use the compact form of the expanded written method to calculate the answer.

Guided practice

$92 \div 23 =$

Estimate: 4

$$
\begin{array}{r}
\boxed{4} \\
\boxed{23} \overline{)\boxed{9}\ \boxed{2}} \\
-\ \boxed{9}\ \boxed{2} \quad \boxed{23} \times \boxed{4} \\
\hline
\boxed{0}
\end{array}
$$

There are 4 groups of 23 in 92.

$92 \div 23 = 4$

Lesson 3: **Dividing 3-digit numbers by 2-digit numbers (1)**

Key words
- dividend
- divisor
- quotient
- expanded written method

- Use the expanded written method to divide 3-digit numbers by 2-digit numbers

Number

Let's learn

$896 \div 28 =$

Estimate by rounding: $900 \div 30 = 30$

```
        3  2
28 ) 8  9  6
  -  8  4  0   (28 × 30)
     ────────
        5  6
  -     5  6   (28 × 2)
     ────────
           0
     ────────
```

$896 \div 28 = 32$

A DIY store has 16 shelves for paint and 768 tins to put on them. How many tins can go on each shelf?

Use the expanded written method to calculate the answer.

Guided practice

$918 \div 34 =$

Estimate: 30

There are 27 groups of 34 in 918.

$918 \div 34 = 27$

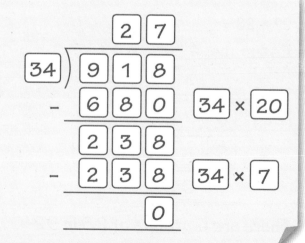

Number

Lesson 4: **Dividing 3-digit numbers by 2-digit numbers (2)**

• Use the long division method to divide 3-digit numbers by 2-digit numbers

Let's learn

396 ÷ 18 =

Estimate by rounding: 400 ÷ 20 = 20

```
         2  2
    ┌──────────
18 )  3  9  6
   –  3  6  ↓
   ─────────
         3  6
   –     3  6
      ─────────
            0
      ─────────
```

396 ÷ 18 = 22

432 people attend a company meeting.

They are asked to sit down at large tables that each have 18 chairs.

If there are just enough seats for everyone at the meeting, how many tables are there in the room?

Guided practice

726 ÷ 33 =

Estimate: 20

There are 22 groups of 33 in 726.

726 ÷ 33 = 22

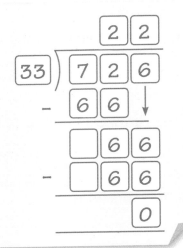

37

Lesson 1: **Decimal place value**

Key words
- decimal
- tenth
- hundredth
- thousandth

- Understand and explain the value of each digit in decimals

Let's learn

The number line between 0·01 and 0·02 can be divided into 10 equal parts.

Each of these 10 equal parts is $\frac{1}{1000}$, one thousandth.

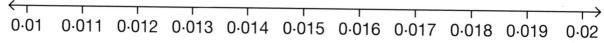

0·01   0·011   0·012   0·013   0·014   0·015   0·016   0·017   0·018   0·019   0·02

You can write any fraction in thousandths as a decimal.

2	3	.	2	5	3
tens	ones		tenths	hundredths	thousandths

LITRES

What is the value of the digits in each amount?

8·237 metres

3·608 kg

0·009 litres

Guided practice

The metal hoop has a mass of 0·12<u>5</u> kg.

What is the value of the underlined digit in the number?

The digit 5 is in the third decimal place.

Using a place value chart, I can see that the 5 has a value of 5 thousandths.

10s	1s	0·1s	0·01s	0·001s
	0	1	2	5

Lesson 2: Composing and decomposing decimals

> **Key words**
> • compose
> • decompose
> • thousandths
> • place value

• Compose and decompose decimals

Let's learn

Decomposing a number means splitting a number into seperate, smaller parts.
One way to do this is to split a number into its separate place values.

$$4·376$$

$$4·376 = 4 + 0·3 + 0·07 + 0·006$$

Composing a number means forming a number from separate, smaller parts.
One example of this is forming a number from separate place values.

7·925 is composed of:

7 ones	7
9 tenths	0·9
2 hundredths	0·02
5 thousandths	+ 0·005
	7·925

👥 Explain how you would decompose the number 6·837 into its separate place values.

Guided practice

Decompose the number 7·234 by place value.

$$7·234 = 7 + 0·2 + 0·03 + 0·004$$

Lesson 3: **Regrouping decimals**

Key words
- decompose
- regroup

- Regroup decimals to help with calculations

Let's learn

Numbers can be split by place value but they can also be decomposed in other ways.

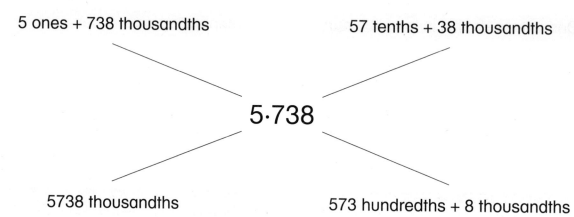

5 ones + 738 thousandths

57 tenths + 38 thousandths

5·738

5738 thousandths

573 hundredths + 8 thousandths

Write each number as a decimal.

  a 9 ones and 927 thousandths

  b 5 tens, 3 ones and 28 thousandths

  c 9 tens, 9 tenths and 9 thousandths

Guided practice

Use the number line to help you regroup each number.

i −17 = $\boxed{-10, -5, -2}$

```
      -2              -5                          -10
  ←—————————————————————————————————————————————————————→
   -17    -15              -10                            0
```

ii −8·54 = $\boxed{-8, -0·5, -0·04}$

```
    -0·04         -0·5                      -8
  ←—————————————————————————————————————————————————————→
   -8·54   -8·5            -8                            0
```

Lesson 4: **Comparing and ordering decimals**

Key words
- compare
- order
- place value
- trailing zero

- Compare and order decimals with two decimal places

Let's learn

To compare numbers, write them in a place value grid, lining up the decimal points.

Which is greater, 2·21 or 2·12?

10s	1s	.	0·1s	0·01s
	2	.	2	1
	2	.	1	2

2·21 > 2·12

Five sprinters run a 100-metre race. Their finishing times are given in the table.

Who was fastest?

Who was slowest?

Who finished in the middle position?

Athlete name	Finishing time
Jacob	13·19
Alexa	13·12
Hassan	12·89
Maria	13·21
Daisy	13·13

Guided practice

Use the symbols < or > to compare the numbers.

a 0·8 $\boxed{>}$ 0·7

b 0·42 $\boxed{<}$ 0·43

Number

Lesson 1: Multiplying whole numbers and decimals by 10, 100 and 1000

Key words
- multiply
- place value

- Multiply whole numbers and decimals by 10, 100 and 1000

Let's learn

100 000s	10 000s	1000s	100s	10s	1s	0·1s	0·01s	
				7	8	2	9	
			7	8	2	9		× 10
		7	8	2	9			× 100
	7	8	2	9	0			× 1000

When multiplied by 10, a number becomes 10 times larger and the digits move 1 place to the left.

When multiplied by 100, a number becomes 100 times larger and the digits move 2 places to the left.

When multiplied by 1000, a number becomes 1000 times larger and the digits move 3 places to the left.

How many different calculations that involve multiplying by 10, 100 or 1000 can you write that will give the answer:

a 673?   b 4500?   c 791 000?

Use your knowledge of place value to convince your partner.

Guided practice

Multiply 19·26 by 10, 100 and 1000.

19·26 × 10   = 192·6   (digits move 1 place to the left)

19·26 × 100   = 1926   (digits move 2 places to the left)

19·26 × 1000 = 19 260 (digits move 3 places to the left)

Lesson 2: **Dividing whole numbers and decimals by 10, 100 and 1000**

> **Key words**
> • divide
> • place value

- Divide whole numbers and decimals by 10, 100 and 1000

Let's learn

10 000s	1000s	100s	10s	1s	0·1s	0·01s	0·001s	
	7	4	0	3				
		7	4	0	3			÷ 10
			7	4	0	3		÷ 100
				7	4	0	3	÷ 1000

When divided by 10, a number becomes 10 times smaller and the digits move 1 place to the right.

When divided by 100, a number becomes 100 times smaller and the digits move 2 places to the right.

When divided by 1000, a number becomes 1000 times smaller and the digits move 3 places to the right.

How many different calculations that involve dividing by 10, 100 or 1000 can you write that will give the answer:

a 4·3? **b** 8·21? **c** 37·639?

Use your knowledge of place value to convince your partner.

> **Guided practice**
> Divide 827 by 10, 100 and 1000.
> 827 ÷ 10 = 82·7 (digits move 1 place to the right)
> 827 ÷ 100 = 8·27 (digits move 2 places to the right)
> 827 ÷ 1000 = 0·827 (digits move 3 places to the right)

Number

Lesson 3: **Rounding decimals to the nearest tenth**

- Round decimals to the nearest tenth

Let's learn

To round decimals to the nearest **tenth**, look at the digit in the hundredths position:

- if it is 5 or greater, round up the tenths digit
- if it is less than 5, the tenths digit remains the same.

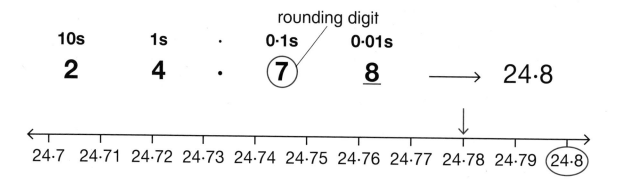

Write all the decimal numbers with two decimal places that round to 16·4 when rounded to the nearest tenth.

Guided practice

Round each number to the nearest tenth.

a 3·53   **b** 11·86   **c** 20·35

a 3·53   This rounds to 3·5 as the hundredths digit is less than 5.

b 11·86   This rounds to 11·9 as the hundredths digit is greater than 5.

c 20·35   This rounds to 20·4 as the hundredths digit is 5.

Lesson 4: **Rounding decimals to the nearest whole number**

Key words
- round
- rounding digit
- place value

Number

- Round decimals to the nearest whole number

Let's learn

To round decimals to the nearest **whole number**, look at the digit in the tenths position:

- if it is 5 or greater, round up the ones digit
- if it is less than 5, the ones digit remains the same.

10s	1s	.	0·1s	0·01s	
4	**②**	**.**	**6**	**2**	⟶ 43

rounding digit

1 Write five 2-place decimals where the value of the tenths digit means you round the number down and the value of hundredths digit means you round the number up.

Round each number to the nearest tenth and then to the nearest whole number.

Guided practice

Round each number to the nearest whole number.

a 7·67   **b** 14·28   **c** 35·55

a 7·67   This rounds to 8 as the tenths digit is greater than 5.

b 14·28   This rounds to 14 as the tenths digit is less than 5.

c 35·55   This rounds to 36 as the tenths digit is 5.

Number

Lesson 1: **Fractions as division**

- Understand that a fraction can be thought of as a division of the numerator by the denominator

Key words
- numerator
- denominator
- decimal

Let's learn

A fraction can be represented as a division of the numerator by the denominator.

$$\frac{3}{4} \rightarrow 3 \div 4$$

The quotient of this division is a decimal.

$$\frac{3}{4} = 0{\cdot}75$$

Two whole cakes and three-quarters of another cake were eaten at a party.

How do you write this amount as a fraction?

How would you represent this fraction as a division?

Guided practice

There are 5 seeds in a packet.

4 out of 5 of the seeds are sunflower seeds.

Write the number of sunflower seeds as a fraction. Express the fraction as a division.

4 out of 5 seeds is represented by the fraction $\frac{4}{5}$.

I can write this fraction as a division: $4 \div 5$.

Lesson 2: **Simplifying fractions**

Number

- Simplify a fraction to its lowest terms

Key words
- **numerator**
- **denominator**
- **highest common factor**
- **simplify**

Let's learn

To simplify a fraction, you divide the numerator and the denominator by the **highest common factor** of both numbers.

Example:

Write the fraction $\frac{16}{40}$ in its simplest form.

Factors of 16: 1, 2, 4, 8, 16

Factors of 40: 1, 2, 4, 5, 8, 10, 20, 40

Common factors of 16 and 40: 1, 2, 4, 8

Highest common factor is 8.

Divide numerator and denominator by 8.

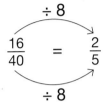

$\frac{16}{40}$ in its simplest form is $\frac{2}{5}$.

7 Zane's teacher asked him to simplify a fraction.

Here is his working:

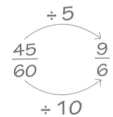

Has Zane made an error? If so, describe it. What would you say to Zane to correct his thinking?

Guided practice

Simplify the fraction $\frac{48}{80}$.

First, find the factors of the numerator and denominator.

48: 1 2 3 4 6 8 12 16 24 48
80: 1 2 4 5 8 10 16 20 40 80

The common factors of 48 and 80 are 1, 2, 4, 8 and 16.

The highest common factor is 16.

Next, divide the numerator and denominator by 16.

$\frac{48}{80}$ in its simplest form is $\frac{3}{5}$.

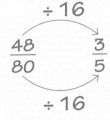

Lesson 3: **Comparing fractions with different denominators**

Key words
- **equivalent fraction**
- **numerator**
- **denominator**

- Compare fractions with different denominators

Let's learn

Here are two methods for comparing fractions with different denominators.

Method 1: Finding equivalent fractions

Which is greater, $\frac{1}{4}$ or $\frac{3}{8}$?

$$\frac{1}{4} \xrightarrow[\times 2]{\times 2} = \frac{2}{8}$$

Therefore $\frac{3}{8} > \frac{1}{4}$.

Method 2: Converting to decimals

Which is greater, $\frac{3}{5}$ or $\frac{1}{2}$?

$\frac{3}{5} = 0.6 \quad \frac{1}{2} = 0.5$

Therefore $\frac{3}{5} > \frac{1}{2}$.

Ruby has three T-shirts. Two of them are blue.

Jamie has 9 T-shirts. Seven of them are blue.

Who has the greater fraction of blue T-shirts? How do you know?

Guided practice

Which is greater, $\frac{4}{5}$ or $\frac{7}{10}$?

Convert one fraction to have the same denominator as the other.
Then compare fractions using < or >.

Convert $\frac{4}{5}$ to a fraction with a denominator of 10.

As $\frac{4}{5}$ is equivalent to $\frac{8}{10}$, $\frac{4}{5} > \frac{7}{10}$.

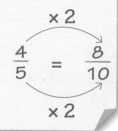

$$\frac{4}{5} \xrightarrow[\times 2]{\times 2} = \frac{8}{10}$$

Lesson 4: **Ordering fractions with different denominators**

Key words
- equivalent fraction
- numerator
- denominator

Number

- Order fractions with different denominators

Let's learn

Order fractions by converting them to the same denominator.

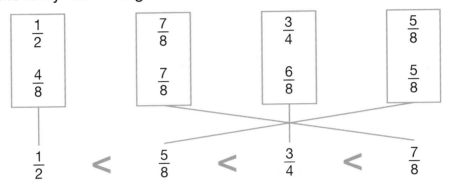

$$\frac{1}{2} \quad < \quad \frac{5}{8} \quad < \quad \frac{3}{4} \quad < \quad \frac{7}{8}$$

Three children play video games.

Jasmine scores 3 points out of 10 in her game.

Luke scores 2 points out of 5 in his game.

Yang scores 1 out of 2 points in her game.

Order the children by the fraction of points they scored, from least to greatest.

Guided practice

Convert the fractions to the same denominator and then order them.

$$\frac{3}{4} \qquad \frac{3}{8} \qquad \frac{1}{4}$$

$$\boxed{\frac{6}{8}} \qquad \boxed{\frac{3}{8}} \qquad \boxed{\frac{2}{8}}$$

$$\frac{1}{4} \quad < \quad \frac{3}{8} \quad < \quad \frac{3}{4}$$

Lesson 1: **Fractions as operators**

- Use a proper or improper fraction as an operator to find the fraction of a quantity

Let's learn

Method 1

$\frac{1}{4}$ of \$80 = \$80 ÷ 4 = \$20

$\frac{7}{4}$ of \$80 = 7 × \$20 = \$140

What is $\frac{7}{4}$ of \$80?

Method 2

$\frac{7}{4}$ of \$80 = 7 × \$80 ÷ 4

= \$560 ÷ 4

= \$140

Lucy manages a small shop.

On Monday, she sells 30 kg of carrots.

On Tuesday, she sells $\frac{8}{5}$ times as many carrots as on Monday.

What amount of carrots were sold on Tuesday?

Guided practice

Jamie has \$35 to spend.

He buys some items in a shop and works out he has spent $\frac{4}{5}$ of his money. How much is that?

First I estimate. $\frac{4}{5}$ of \$35 will be less than 35, around 30.

I can find $\frac{1}{5}$ of \$35 by dividing by 5.   \$35 ÷ 5 = \$7

I find $\frac{4}{5}$ of \$35 by multiplying $\frac{1}{5}$ of \$35 by 4.   \$7 × 4 = \$28.

$\frac{4}{5}$ of \$35 is \$28.

Lesson 2: **Adding and subtracting fractions**

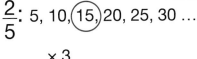

Key words
- proper fraction
- improper fraction
- denominator
- lowest common multiple (LCM)
- common denominator

- Add and subtract fractions with different denominators

Let's learn

$$\frac{1}{3} + \frac{2}{5} =$$

Finding a common denominator

$\frac{1}{3}$: 3, 6, 9, 12, ⑮, 18 …

$\frac{2}{5}$: 5, 10, ⑮, 20, 25, 30 …

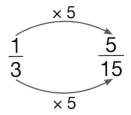

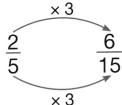

$$\frac{1}{3} + \frac{2}{5} = \frac{5}{15} + \frac{6}{15} = \frac{11}{15}$$

Akihiro spends $\frac{2}{7}$ of the money she received from her parents on a book.
She also spends $\frac{2}{5}$ of the money on a painting set.
What fraction of the money has Akihiro spent in total?

Guided practice

$$\frac{9}{10} - \frac{2}{3} =$$

First I estimate. $\frac{9}{10}$ is nearly 1. 1 subtract $\frac{2}{3}$ is $\frac{1}{3}$.
So the answer will be close to $\frac{1}{3}$.

To subtract the fractions, I need to convert them to fractions with the same denominator.

First, I work out the common denominator.

The lowest common multiple of 10 and 3 is 30.

I convert each fraction to a fraction with a denominator of 30.

$$\frac{9}{10} - \frac{2}{3} = \frac{27}{30} - \frac{20}{30} = \frac{7}{30} \qquad \frac{9}{10} - \frac{2}{3} = \frac{7}{30}$$

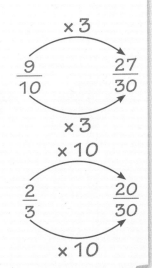

Number

Lesson 3: **Multiplying fractions by whole numbers**

Key words
• proper fraction
• product

• Multiply proper fractions by whole numbers

Let's learn

$3 \times \dfrac{5}{6} =$

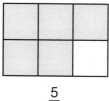   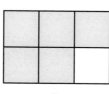   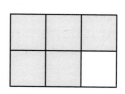

$\dfrac{5}{6}$ + $\dfrac{5}{6}$ + $\dfrac{5}{6}$ $= \dfrac{15}{6}$

$$\dfrac{5}{6} \times 3 = \dfrac{15}{6} \text{ or } 2\dfrac{3}{6} \text{ or } 2\dfrac{1}{2}$$

👥 Use a convincing argument to prove that:

4 slices that are each three-eighths of a pizza are equivalent to $1\dfrac{1}{2}$ whole pizzas. All the pizzas are the same size.

Guided practice

$\dfrac{3}{7} \times 5 =$

I estimate first. $\dfrac{3}{7}$ is nearly $\dfrac{1}{2}$. $5 \times \dfrac{1}{2} = 2\dfrac{1}{2}$

I expect the answer to be close to $2\dfrac{1}{2}$.

I draw a circle divided into 7 equal pieces and shade 3 of them to model the fraction $\dfrac{3}{7}$.

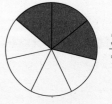
$\dfrac{3}{7}$

Then I draw five of these models and add the fractions together.

         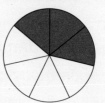 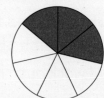

$\dfrac{3}{7}$ + $\dfrac{3}{7}$ + $\dfrac{3}{7}$ + $\dfrac{3}{7}$ + $\dfrac{3}{7}$ $= \dfrac{15}{7}$

I can convert $\dfrac{15}{7}$ to a mixed number: $2\dfrac{1}{7}$.

$\dfrac{3}{7} \times 5 = 2\dfrac{1}{7}$

Lesson 4: **Dividing fractions by whole numbers**

Key words
• **proper fraction**
• **quotient**

• Divide proper fractions by whole numbers

Let's learn

$\frac{7}{8} \div 5 =$

Draw an area model for $\frac{7}{8}$.

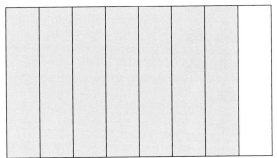

The diagram is now divided into 40 equal parts.

Indicate $\frac{1}{5}$ on the diagram.

One-fifth of $\frac{7}{8}$ is $\frac{7}{40}$

$\frac{7}{8} \div 5 = \frac{7}{40}$

Divide the area model into 5 equal parts to divide by 5.

$\frac{1}{40}$	$\frac{1}{40}$	$\frac{1}{40}$	$\frac{1}{40}$	$\frac{1}{40}$	$\frac{1}{40}$	$\frac{1}{40}$	$\frac{1}{40}$
$\frac{1}{40}$	$\frac{1}{40}$	$\frac{1}{40}$	$\frac{1}{40}$	$\frac{1}{40}$	$\frac{1}{40}$	$\frac{1}{40}$	$\frac{1}{40}$
$\frac{1}{40}$	$\frac{1}{40}$	$\frac{1}{40}$	$\frac{1}{40}$	$\frac{1}{40}$	$\frac{1}{40}$	$\frac{1}{40}$	$\frac{1}{40}$
$\frac{1}{40}$	$\frac{1}{40}$	$\frac{1}{40}$	$\frac{1}{40}$	$\frac{1}{40}$	$\frac{1}{40}$	$\frac{1}{40}$	$\frac{1}{40}$
$\frac{1}{40}$	$\frac{1}{40}$	$\frac{1}{40}$	$\frac{1}{40}$	$\frac{1}{40}$	$\frac{1}{40}$	$\frac{1}{40}$	$\frac{1}{40}$

$\frac{1}{40}$	$\frac{1}{40}$	$\frac{1}{40}$	$\frac{1}{40}$	$\frac{1}{40}$	$\frac{1}{40}$	$\frac{1}{40}$	$\frac{1}{40}$
$\frac{1}{40}$	$\frac{1}{40}$	$\frac{1}{40}$	$\frac{1}{40}$	$\frac{1}{40}$	$\frac{1}{40}$	$\frac{1}{40}$	$\frac{1}{40}$
$\frac{1}{40}$	$\frac{1}{40}$	$\frac{1}{40}$	$\frac{1}{40}$	$\frac{1}{40}$	$\frac{1}{40}$	$\frac{1}{40}$	$\frac{1}{40}$
$\frac{1}{40}$	$\frac{1}{40}$	$\frac{1}{40}$	$\frac{1}{40}$	$\frac{1}{40}$	$\frac{1}{40}$	$\frac{1}{40}$	$\frac{1}{40}$
$\frac{1}{40}$	$\frac{1}{40}$	$\frac{1}{40}$	$\frac{1}{40}$	$\frac{1}{40}$	$\frac{1}{40}$	$\frac{1}{40}$	$\frac{1}{40}$

Prove that:

4 If $\frac{5}{6}$ of a pizza is divided into 4 equal slices, then each slice represents $\frac{5}{24}$ of the whole pizza.

Guided practice

$\frac{3}{5} \div 4 =$

Use the model to divide.

First I estimate. $\frac{3}{5}$ is close to $\frac{1}{2}$.

$\frac{1}{2}$ divided by 4 will be $\frac{1}{8}$. I expect the answer to be close to $\frac{1}{8}$.

I can see that dividing fifths by 4 makes twentieths.

I can also see that one quarter of $\frac{3}{5}$ is $\frac{3}{20}$.

$\frac{3}{5} \div 4 = \frac{3}{20}$

$\frac{1}{20}$	$\frac{1}{20}$	$\frac{1}{20}$	$\frac{1}{20}$
$\frac{1}{20}$	$\frac{1}{20}$	$\frac{1}{20}$	$\frac{1}{20}$
$\frac{1}{20}$	$\frac{1}{20}$	$\frac{1}{20}$	$\frac{1}{20}$
$\frac{1}{20}$	$\frac{1}{20}$	$\frac{1}{20}$	$\frac{1}{20}$
$\frac{1}{20}$	$\frac{1}{20}$	$\frac{1}{20}$	$\frac{1}{20}$

Number

Lesson 1: **Percentages of shapes**

Key words
• per cent (%)
• percentage
• fraction
• decimal

• Recognise percentages of shapes

Let's learn

In the picture, there are 100 cones.

To represent 35%, you shade 35 of the cones.

35 cones are red and the rest are blue.

The number of red cones can be described as a fraction of the total number of cones.

This is $\frac{35}{100}$, which can be simplified by dividing both the numerator and denominator by 5 to make $\frac{7}{20}$.

👥 Using a 100 grid, shade squares on the grid to show 85%.
What fraction is represented by the shaded squares?

Guided practice

Shade the squares in the grid to show 45%.

45% equals 4 × 10% + 5%

To show 45%, I shade 4 columns of the grid plus a half of one column.

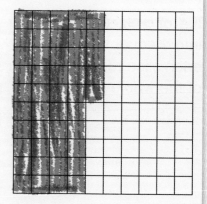

Lesson 2: **Percentages of whole numbers (1)**

> **Key words**
> • percentage
> • fraction
> • denominator

• Calculate percentages of whole numbers and quantities

Let's learn

A percentage of a number or quantity can be calculated in different ways.

15% of 40 g

15% of 40 g
$= (10\%$ of $40\,g) + (5\%$ of $40\,g)$
10% of 40 g $= 40\,g \div 10$
$= 4\,g$
5% of 40 g $= \frac{1}{2}$ of 10% of 40 g
$= 4\,g \div 2$
$= 2\,g$
15% of 40 g $= 4\,g + 2\,g$
$= 6\,g$

45% of 40 g

$45\% = 50\% - 5\%$
50% of 40 g $= 40\,g \div 2$
$= 20\,g$
5% of 40 g $= \frac{1}{20} \times 40\,g$
$= 2\,g$
45% of 40 g $= 20\,g - 2\,g$
$= 18\,g$

75% of 40 g

$75\% = \frac{3}{4}$
$\frac{3}{4}$ of 40 g $= \frac{3}{4} \times 40$
$= 3 \times 40 \div 4$
$= 3 \times 10$
$= 30\,g$

• Daisy has $6 to spend.

If she spends 35% of this amount, how much money does she have left to spend?

Guided practice

65% of $30 =

$65\% = 50\% + 10\% + 5\%$

50% of $30 $= \frac{1}{2} \times 30 = \$15$

10% of $30 $= \$30 \div 10$
$= \$3$

5% of $30 $= \frac{1}{2}$ of 10% of $30
$= \$3 \div 2 = \$1.50$

65% of $30 $= 50\% + 10\% + 5\%$
$= \$15 + \$3 + \$1.50$
$= \$19.50$

Lesson 3: **Percentages of whole numbers (2)**

* Calculate percentages of whole numbers and quantities

Let's learn

What is the new price of the laptop after a 20% discount?

A **discount** is the amount taken off the usual price.

$20\% = \frac{1}{5}$

$\frac{1}{5}$ of \$800 = \$800 ÷ 5 = \$160

ORIGINAL PRICE: \$800

New price:

original price – discount

= \$800 – \$160 = \$640

What is the new price of the car after the price rise?

\$4850

Guided practice

Yesterday the price of a remote-controlled car was \$80.

Today, the shopkeeper has discounted the price by 25%.
What is the new price of the car?

Discount is 25% or $\frac{1}{4}$

$\frac{1}{4}$ of \$80 = \$20

New price: \$80 – \$20 = \$60

Lesson 4: **Comparing percentages**

Key words
* percentage
* quantity

* Compare and order percentages of quantities

Let's learn

Fractions, decimals and percentages can have equivalent values

Convert fractions, decimals and percents

Fractions $\frac{3}{4}$ → Divide numerator by denominator → Decimals 0·75 → Multiply by 100 → Percents 75%

Fractions $\frac{3}{4}$ ← Write as fraction and simplify ← Decimals 0·75 ← Divide by 100 ← Percents 75%

Fraction	Decimal	Percentage
$\frac{1}{2}$	0·5	50%
$\frac{1}{4}$	0·25	25%
$\frac{3}{4}$	0·75	75%
$\frac{1}{5}$	0·2	20%
$\frac{1}{10}$	0·1	10%
$\frac{1}{20}$	0·05	5%
$\frac{4}{10}$	0·4	40%
$\frac{6}{10}$	0·6	60%
$\frac{7}{10}$	0·7	70%
$\frac{8}{10}$	0·8	80%
$\frac{9}{10}$	0·9	90%

We can use the equivalence to order fractions, decimals and percentages.

Example: Order the following numbers:
$\frac{3}{10}$, 0·25, 40%

It is easier to compare the numbers if we convert them to the same form, for example percentages.

$\frac{3}{10} = 30\%$ $0·25 = 25\%$ 40%

As 25% < 30% < 40%, the order will be 0·25, $\frac{3}{10}$, 40%.

Which is greater: 75%, 0·8 or $\frac{7}{10}$? How do you know?

Guided practice

Order the children by the distance they ran, from shortest to longest. They all ran part of 1 km.

Sienna ran 0·6 km.   Robbie ran $\frac{1}{2}$ km.   Maisie ran 70% of 1 km.

I convert the distances to the same form: decimals.

Sienna ran 0·6 km.   Robbie ran 0·5 km.   Maisie ran 0·7 km.

Therefore the order is Robbie (0·5 km), Sienna (0·6 km), Maisie (0·7 km).

Number

Lesson 1: Adding decimals (mental strategies)

• Add pairs of decimals mentally

Let's learn

You can use mental strategies to add decimals with different numbers of decimal places.

$6 \cdot 5 + 0 \cdot 78 =$

a Using place value counters

Step 1

1s	0·1s	0·01s

Step 2

1s	0·1s	0·01s

Step 3

1s	0·1s	0·01s

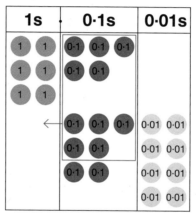

b Using a number line (counting on)

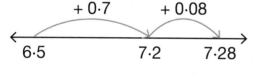

$+ 0 \cdot 7$ $+ 0 \cdot 08$

6·5   7·2   7·28

c Using a compensation strategy

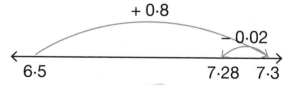

$+ 0 \cdot 8$ $- 0 \cdot 02$

6·5   7·28   7·3

👥 What is the combined mass of the two boxes?

Which method did you use to work out the answer?

4·5 kg   0·88 kg

Guided practice

Add the numbers mentally using any strategy you prefer.

$5 \cdot 8 + 0 \cdot 43 = \boxed{6 \cdot 23}$

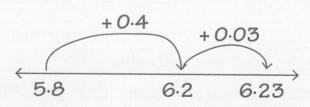

$+ 0 \cdot 4$ $+ 0 \cdot 03$

5·8   6·2   6·23

Lesson 2: **Adding decimals (written methods)**

Key words
- place value
- regroup
- addend
- trailing zero

Number

- Add pairs of decimals using written methods

Let's learn

Use written strategies when numbers are too large to add mentally.

$8·678 + 76·58 =$

Estimate first by rounding: $77 + 9 = 86$

Expanded written method	**Formal written method**

Expanded written method

```
    8·678
 + 76·58
  00·008
  00·150
  01·100
  14·000
  70·000
  85·258
```

Formal written method

```
    8·678
 + 76·58
  85·258
   1 1 1
```

Two amounts of sand, one large and one small, are poured into each of three containers, A, B and C.

Container A: 3·658 kg and 67·27 kg

Container B: 65·78 kg and 5·546 kg

Container C: 4·966 kg and 66·86 kg

Find the total amount of sand in each container and order the containers by the mass of sand they contain, from least to greatest.

Guided practice

Calculate, using the formal written method. Estimate the answer first.

$8·868 + 57·77 =$

Estimate: $\boxed{58 + 9 = 67}$

```
    8·868
 + 57·770
  66·638
   1 1 1
```

Lesson 3: **Subtracting decimals (mental strategies)**

Number

• Subtract pairs of decimals mentally

Let's learn

Mental strategies for subtraction

$7.7 - 2.57 =$

a Using place value counters

Step 1

1s	0·1s	0·01s
1 1	0·1 0·1 0·1	
1 1	0·1 0·1 0·1	
1 1	0·1	
1		

Step 2

1s	0·1s	0·01s
1 1	0·1 0·1 0·1	0·01 0·01 0·01
1 1	0·1 0·1 0·1	0·01 0·01 0·01
1 1	– 0·5	0·01 0·01 0·01
1 – 2		0·01
		– 0·07

Step 3

1s	0·1s	0·01s
1 1	0·1	0·01 0·01 0·01
1 1		
1		

b Counting back

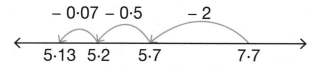

$$-0.07 \quad -0.5 \quad -2$$

5·13   5·2   5·7   7·7

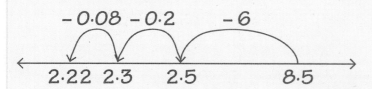
Adil has $9·40 and spends $4·17.

How much money does Adil have left?

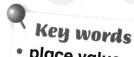

Number

Lesson 4: **Subtracting decimals (written methods)**

Key words
- place value
- regroup
- trailing zero

- Subtract pairs of decimals using written methods

Let's learn

74·375 − 29·85 =

Estimate first by rounding: 74 − 30 = 44

Formal written method

$$\begin{array}{r} {}^6\cancel{7}\,{}^{13}\cancel{4}\cdot{}^1 3\ 7\ 5 \\ -\ 2\ 9\cdot 8\ 5\ 0 \\ \hline 4\ 4\cdot 5\ 2\ 5 \end{array}$$

Emily has three large containers of water. She pours out a small amount from each container. Work out the missing amounts in the table below.

Container	Original amount (*l*)	Amount poured (*l*)	Amount left (*l*)
A	8·4	5·17	
B	12·334	8·71	
C	15·7		12·32

Guided practice

Calculate, using the formal written method. Estimate the answer first.

9·7 − 6·46 =

Estimate: 9·7 − 6·5 = 3·2

$$\begin{array}{r} 9\cdot{}^6\cancel{7}\,{}^1 0 \\ -\ 6\cdot 4\ 6 \\ \hline 3\cdot 2\ 4 \end{array}$$

61

Lesson 1: **Multiplying decimals by 1-digit whole numbers (1)**

Key words
- place value
- grid method
- product
- partial product

- Multiply decimals by 1-digit whole numbers

Let's learn

$235 \cdot 4 \times 6 =$

Place value counters

×	200	30	5	0·4
6	100 100 / 100 100 / 100 100 / 100 100 / 100 100 / 100 100	10 10 10 / 10 10 10 / 10 10 10 / 10 10 10 / 10 10 10 / 10 10 10	1 1 1 1 1 / 1 1 1 1 1 / 1 1 1 1 1 / 1 1 1 1 1 / 1 1 1 1 1 / 1 1 1 1 1	0·1 0·1 0·1 0·1 / 0·1 0·1 0·1 0·1 / 0·1 0·1 0·1 0·1 / 0·1 0·1 0·1 0·1 / 0·1 0·1 0·1 0·1 / 0·1 0·1 0·1 0·1

Grid method

×	200	30	5	0·4
6	1200	180	30	2·4

$= 1412 \cdot 4$

An apple has a mass of 267·8 g.

What is the combined mass of 7 identical apples?

267·8 g

Guided practice

Use the grid method to multiply. Estimate the answer first.

$473 \cdot 6 \times 6 =$ | 2841·6 |

Estimate: | $500 \times 6 = 3000$ |

×	400	70	3	0·6
6	2400	420	18	3·6

I find the sum of the partial products:

$2400 + 420 + 18 + 3 \cdot 6 = 2841 \cdot 6$

$473 \cdot 6 \times 6 = 2841 \cdot 6$

Lesson 2: **Multiplying decimals by 1-digit whole numbers (2)**

• Multiply decimals by 1-digit whole numbers

Let's learn

$453 \cdot 8 \times 6 =$

Partitioning

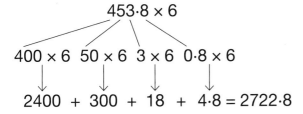

$453 \cdot 8 \times 6$

$400 \times 6 \quad 50 \times 6 \quad 3 \times 6 \quad 0 \cdot 8 \times 6$

$2400 + 300 + 18 + 4 \cdot 8 = 2722 \cdot 8$

Short multiplication

$453 \cdot 8 \times 6$ is equivalent to $4538 \times 6 \div 10$

```
    4  5  3  8
×          6
 2  7  2  2  8
    3  2  4
```

$27\,228 \div 10 = 2722 \cdot 8$

👥 Each tyre has a mass of $46 \cdot 87$ kg.
What is the combined mass of 6 identical tyres?

46·87 kg

Expanded written method

$453 \cdot 8 \times 6$ is equivalent to $4538 \times 6 \div 10$

```
    4  5  3  8
×          6
       4  8   (8 × 6)
    1  8  0   (30 × 6)
 3  0  0  0   (500 × 6)
 2  4  0  0  0   (4000 × 6)
 2  7  2  2  8
       1
```

$27\,228 \div 10 = 2722 \cdot 8$

Guided practice

Use the expanded written method to multiply. Estimate the answer first.

$32 \cdot 48 \times 7 =$

Estimate: $30 \times 7 = 210$

$32 \cdot 48 \times 7 = 3248 \times 7 \div 100$

```
    3  2  4  8
×          7
       5  6   (8 × 7)
    2  8  0   (40 × 7)
 1  4  0  0   (200 × 7)
 2  1  0  0  0   (3000 × 7)
 2  2  7  3  6
       1
```

$22\,736 \div 100 = 227 \cdot 36$

$32 \cdot 48 \times 7 = 227 \cdot 36$

Lesson 3: **Multiplying decimals by 2-digit whole numbers (1)**

• Multiply decimals by 2-digit whole numbers

Let's learn

$48.6 \times 37 =$

Estimate: $50 \times 40 = 2000$

Grid method

×	40	8	0.6
30	1200	240	18
7	280	56	4.2

```
  1458.0
+  340.2
  1798.2
```

$48.6 \times 37 = 1798.2$

A mountaineer climbs 2.87 metres every minute.

How far will she have climbed after 48 minutes?

Guided practice

Use the grid method to multiply. Estimate the answer first.

$948.3 \times 37 =$ | 35 087.1 |

Estimate: | $900 \times 40 = 36 000$ |

×	900	40	8	0.3
30	27 000	1200	240	9
7	6300	280	56	2.1

```
   28449
+  6638.1
   35087.1
    1 1  1
```

Lesson 4: **Multiplying decimals by 2-digit whole numbers (2)**

Key words
- place value
- partition
- expanded written method
- long multiplication

Number

- Multiply decimals by 2-digit whole numbers

Let's learn

$36 \cdot 36 \times 27 =$

Partitioning

$36 \cdot 36 \times 27 = (30 \times 27) + (6 \times 27) + (0 \cdot 3 \times 27) + (0 \cdot 06 \times 27)$
$\qquad\qquad = 810 + 162 + 8 \cdot 1 + 1 \cdot 62$
$\qquad\qquad = 981 \cdot 72$

Expanded written method

$36 \cdot 36 \times 27$ is equivalent to
$3636 \times 27 \div 100$

```
      3  6  3  6
×           2  7
   2  5  4  5  2   (3636 × 7)
   7  2  7  2  0   (3636 × 20)
   9  8  1  7  2
         1
```

$98\,172 \div 100 = 981 \cdot 72$

Long multiplication

$36 \cdot 36 \times 27$ is equivalent to
$3636 \times 27 \div 100$

```
      3  6  3  6
×           2  7
   2  5 ⁴4 ²5 ⁴2
   7 ¹2  7 ¹2  0
   9  8  1  7  2
         1
```

$98\,172 \div 100 = 981 \cdot 72$

A bird flies at an altitude of $17 \cdot 67 \, \text{m}$.

What would be the bird's altitude if it flew 33 times higher?

Guided practice

Use the expanded written method to multiply. Estimate the answer first.

$576 \cdot 4 \times 38 =$

Estimate: $600 \times 40 = 24\,000$

$576 \cdot 4 \times 38 = 5764 \times 38 \div 10$

```
         5  7  6  4
×              3  8
      4  6  1  1  2   (5764 × 8)
   1  7  2  9  2  0   (5764 × 30)
   2  1  9  0  3  2
      1     1
```

$21\,9032 \div 10 = 21\,903 \cdot 2$

$576 \cdot 4 \times 38 = 21\,903 \cdot 2$

Number

Lesson 1: **Dividing one-place decimals by whole numbers (1)**

Key words
- partition
- expanded written method

- Divide one-place decimals by whole numbers

Let's learn

$36.6 \div 6 =$

Estimate by rounding: $36 \div 6 = 6$

Partitioning (mental method)

$36.6 = 30 + 6 + 0.6$

[Divide using the distributive property of number]

$36.6 \div 6 = (30 \div 6) + (6 \div 6) + (0.6 \div 6)$
$\qquad\quad = 5 + 1 + 0.1$
$\qquad\quad = 6.1$

or

[Divide using the distributive property of number]

$36.6 \div 6 = (36 \div 6) + (0.6 \div 6)$
$\qquad\quad = 6.1$

$147.6 \div 6 =$

Expanded written method

$147.6 \div 6$ is equivalent to $1476 \div 6 \div 10$

```
      246
6 ) 1476
  − 1200
      276
  −   240
       36
  −    36
        0
```

$246 \div 10 = 24.6$

Therefore $147.6 \div 6 = 24.6$

Three identical chairs have a combined mass of 222.6 kg. What is the mass of each chair?

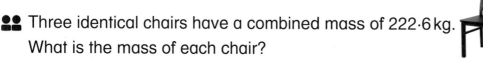

Guided practice

Estimate first and then use the expanded written method of division to work out the answer to the calculation.

$86.6 \div 4 =$ | 21.65 | Estimate: | 20 |

$86.6 \div 4$ is equivalent to | $866 \div 4 \div 10$ |

```
       216½ = 216.5
4 ) 866
  − 800
     66
  −  40
     26
  −  24
      2
```

| 216.5 | $\div 10 =$ | 21.65 |

Number

Lesson 2: **Dividing one-place decimals by whole numbers (2)**

Key word
• short division

• Divide one-place decimals by whole numbers

Let's learn

$354 \cdot 8 \div 4 =$

Estimate by rounding: $360 \div 4 = 90$

Short division

$354 \cdot 8 \div 4$ is equivalent to $3548 \div 4 \div 10$

$$4\overline{)3^3 5^3 4^2 8} = 887$$

$887 \div 10 = 88 \cdot 7$

Therefore, $354 \cdot 8 \div 4 = 88 \cdot 7$

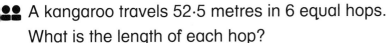 A kangaroo travels $52 \cdot 5$ metres in 6 equal hops. What is the length of each hop?

Guided practice

Estimate first and then use the short division method to work out the answer to the calculation.

$265 \cdot 2 \div 3 = \boxed{88.4}$ Estimate: $\boxed{90}$ $3\overline{)2^2 6^2 5^1 2} = 884$

$265 \cdot 2 \div 3$ is equivalent to $\boxed{2652 \div 3 \div 10}$ $\boxed{884} \div 10 = \boxed{88.4}$

Number

Lesson 3: **Dividing two-place decimals by whole numbers (1)**

Key words
* partition
* expanded written method

• Divide two-place decimals by whole numbers

Let's learn

96·39 ÷ 3 =

Estimate by rounding: 90 ÷ 3 = 30

Partitioning (mental method)

96·39 = 96 + 0·39

[Divide using the distributive property of number]

96·39 ÷ 3 = (96 ÷ 3) + (0·39 ÷ 3)
= 32 + 0·13
= 32·13

28·12 ÷ 4 =

Expanded written method

28·12 ÷ 4 is equivalent to 2812 ÷ 4 ÷ 100

$$\begin{array}{r} 703 \\ 4\overline{)2812} \\ -\ 2800 \\ \hline 12 \\ -\ 12 \\ \hline 0 \end{array}$$

703 ÷ 100 = 7·03

Therefore 28·12 ÷ 4 = 7·03.

Mr Jones shares $96·84 equally between 12 people in his family.

How much does each person receive?

Guided practice

Estimate first and then use the expanded written method of division to work out the answer to the calculation.

42·54 ÷ 6 = $\boxed{7.09}$

Estimate: $\boxed{7}$

42·54 ÷ 6 is equivalent to

$\boxed{4254 \div 6 \div 100}$

$$\begin{array}{r} 709 \\ 6\overline{)4254} \\ -\ 4200 \\ \hline 54 \\ -\ 54 \\ \hline 0 \end{array}$$

$\boxed{709}$ ÷ 100 = $\boxed{7.09}$

Lesson 4: **Dividing two-place decimals by whole numbers (2)**

Key word
• short division

• Divide two-place decimals by whole numbers

Let's learn

49·35 ÷ 7 =

Estimate by rounding: 49 ÷ 7 = 7

Short division

49·35 ÷ 7 is equivalent to 4935 ÷ 7 ÷ 100.

$$\begin{array}{r} 7\ 0\ 5 \\ 7\overline{)4\,^49\ 3\,^35} \end{array}$$

705 ÷ 100 = 7·05

Therefore, 49·35 ÷ 7 = 7·05.

Billy cuts 40·56 metres of ribbon into 8 equal pieces.
What is the length of each piece?

Guided practice

Estimate first and then use the short division method to work out the answer to the calculation.

54·27 ÷ 9 = $\boxed{6.03}$ Estimate: $\boxed{6}$

$$\begin{array}{r} 6\ 0\ 3 \\ 9\overline{)5\,^54\ 2\,^27} \end{array}$$

54·27 ÷ 9 is equivalent to $\boxed{5427 \div 9 \div 100}$ $\boxed{603}$ ÷ 100 = $\boxed{6.03}$

Lesson 1: **Direct proportion (1)**

Key words
- proportion
- in every
- direct proportion
- scale (scaling)
- scale factor

- Understand the relationship between two quantities when they are in direct proportion

Let's learn

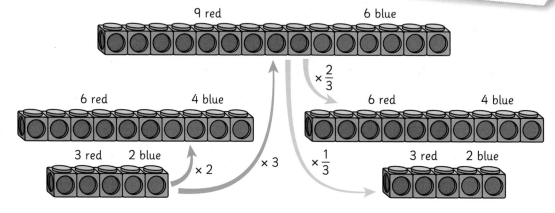

The **proportion** of red cubes is $\frac{3}{5}$. The proportion of blue cubes is $\frac{2}{5}$.

Double the number of cubes to 10 cubes. The number of cubes is now 2 times larger.

The cubes are still in proportion: $\frac{6}{10}$ ($\frac{3}{5}$) red and $\frac{4}{10}$ ($\frac{2}{5}$) blue.

Multiply the number of cubes by 3, to 15 cubes. The number of cubes is now 3 times larger.

The cubes are still in proportion: $\frac{9}{15}$ ($\frac{3}{5}$) red and $\frac{6}{15}$ ($\frac{2}{5}$) blue.

You can also reduce the number of cubes in proportion by dividing by a divisor or multiplying by a fraction.

For example, dividing by 3 or multiplying by a $\frac{1}{3}$ would make the number of cubes 3 times smaller.

A model of a dog is in proportion to a real dog.

The dog is 9 times larger than the model.

If the model is 4 cm tall, the real dog will be 36 cm tall ($9 \times 4 = 36$).

If the real dog's tail is 18 cm long, how long will the tail on the model be?

Guided practice

Two toy rockets, rocket A and rocket B, are in proportion.

Rocket B is 5 times larger than rocket A.

If rocket B is 90 cm tall, how tall is rocket A?

As the rockets are in proportion, and rocket B is 5 times larger than rocket A, then rocket A must be 5 times smaller.

Height of rocket B = 90 cm

Height of rocket A = 90 cm ÷ 5 = 18 cm

Lesson 2: **Direct proportion (2)**

- Understand the relationship between two quantities when they are in direct proportion

Let's learn

Multiplying by a number greater than 1 increases a quantity or measure in proportion.

The side lengths of the triangle are now 2 times larger.

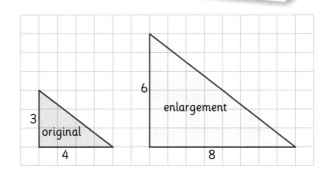

Multiplying by a number less than 1 decreases a quantity or measure in proportion.

The quantity of each ingredient is now 4 times smaller.

for a quarter of the number of pancakes

multiply by $\frac{1}{4}$ or divide by 4

Fred completes 16 pieces of homework every 2 months.

How many pieces of homework will Fred complete in 8·5 months?

Assume the relationship is directly proportional.

Guided practice

Hannah buys 6 cans of lemonade for $8.

How many cans can she buy for:

a $16?   **b** $32?   **c** $4?

As $16 is double $8, she can buy 12 cans for $32. (2 × 6 = 12)

As $32 is four times $8, she can buy 24 cans for $32. (4 × 6 = 24)

As $4 is half $8, she can buy 3 cans for $4. (6 ÷ 2 = 3)

Number

Lesson 3: **Equivalent ratios (1)**

• Use equivalent ratios to calculate unknown amounts

Key words
• ratio
• equivalent ratio
• for every
• to every

Let's learn

These ratios are all equivalent.

$$\times 5 \begin{pmatrix} 2 & : & 6 \end{pmatrix} \times 5$$
$$10 & : & 30$$
$$\div 10 \begin{pmatrix} & & \end{pmatrix} \div 10$$
$$1 & : & 3$$
$$\times 3 \begin{pmatrix} & & \end{pmatrix} \times 3$$
$$3 & : & 9$$
$$\times 2 \begin{pmatrix} & & \end{pmatrix} \times 2$$
$$6 & : & 18$$

The ratio of children wearing shoes to trainers is 8 for every 3.

How many children will be wearing shoes if 51 children are wearing trainers?

Guided practice

Draw a ring around the ratios that are equivalent to the ratio in the box.

| 7:3 | 14:6 | 23:12 | 70:30 | 35:15 | 21:9 | 9:21 |

$7:3 \xrightarrow{\times 2} 14:6 ✓$ $7:3 \xrightarrow{\times 4} 28:12 ✗$ $7:3 \xrightarrow{\times 10} 70:30 ✓$

not 23:12

$7:3 \xrightarrow{\times 5} 35:15 ✓$ $7:3 \xrightarrow{\times 3} 21:9 ✓$ $7:3 \xrightarrow{\times 7} 49:21 ✗$

not 9:21

Lesson 4: **Equivalent ratios (2)**

- Use equivalent ratios to calculate unknown amounts

Key words
- ratio
- equivalent ratio
- for every
- to every

Let's learn

Tables are useful when comparing equivalent ratios.

Grape soda recipe

Mix cups of grape juice and soda water in the ratio 3:2.

grape juice	3	6	12	24
soda water	2	4	8	16

In the table, the ratio is doubled each time so the ingredients are kept in proportion.

The ratio of chocolate to banana muffins on trays is 7:4.

If there are 16 banana muffins, how many chocolate muffins are there?

Guided practice

To make light blue paint, Penny uses 7 drops of white paint for every 2 drops of blue paint.

a If she uses 12 drops of blue paint, how many drops of white paint will she need?

b If she uses 49 drops of white paint, how many drops of blue paint will she need?

Changing the amount of blue paint from 2 drops to 12 drops is an increase of 6 times the amount (12 ÷ 2 = 6).

To keep proportion, increase the amount of white paint by 6 times: 7 × 6 = 42.

Changing the amount of white paint from 7 drops to 49 drops is an increase of 7 times the amount (49 ÷ 7 = 7).

To keep proportion, increase the amount of blue paint by 7 times: 2 × 7 = 14.

white paint (drops)	7	42	49
blue paint (drops)	2	12	14

Lesson 1: **Quadrilaterals**

- Identify, describe and sketch quadrilaterals

Let's learn

The table below describes some of the basic properties of each quadrilateral.

Quadrilateral	Properties	
rectangle or oblong	4 right angles and opposite sides equal	
square	4 right angles and 4 equal sides	
parallelogram	two pairs of parallel sides and opposite sides equal	
rhombus	parallelogram with 4 equal sides	
trapezium or trapezoid	two sides parallel	
kite	two pairs of adjacent sides of the same length	

5 Characterise and describe two differences and two similarities between a rhombus and a rectangle.

rhombus   rectangle

Guided practice

Write yes or no to answer each question. You may need to write a short explanation to answer some of the questions.

Property	Kite
Opposite sides are equal?	no
Opposite sides are parallel?	no
Adjacent sides are equal?	yes two pairs
All angles are equal 90°?	no
Opposite angles are equal?	yes one pair only
Diagonals bisect?	yes one pair only
Lines of symmetry?	yes 1 line

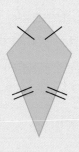

Key words
- quadrilateral
- congruent
- rectangle
- square
- rhombus
- parallelogram
- kite
- trapezium
- diagonal
- parallel
- perpendicular
- bisect

Geometry and Measure

Lesson 2: **Parts of a circle**

Key words
• **circle**
• **centre**
• **radius/radii**
• **diameter**
• **circumference**

• Identify and label parts of a circle

Let's learn

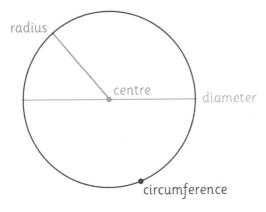

The diameter d is twice the radius r.

$$d = 2 \times r$$

Example: The diameter of a circle is 72 cm. What is the radius?

$2 \times r = d$

$r = d \div 2$ (or $\frac{1}{2} \times d$) (using the inverse operation of division)

$r = 72 \div 2 = 36$

The radius of the circle is 36 cm.

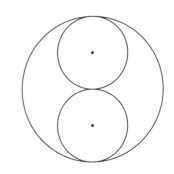

The diagram shows two circles aligned vertically through the centres.

If the radius of each small circle is 15 cm, what is the diameter of the large circle?

Guided practice

Mark and label one radius and one diameter on the circle.

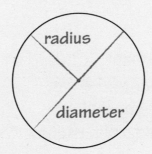

radius

diameter

Geometry and Measure

Lesson 3: **Constructing circles**

- Construct circles of a given radius or diameter

> **Key words**
> - circle
> - centre
> - radius/radii
> - diameter
> - circumference
> - compass
> - separation distance

Let's learn

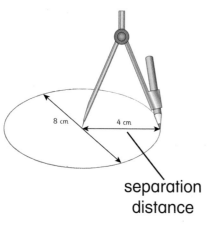

separation distance

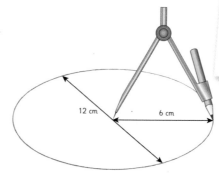

To draw a circle with a radius of 4 cm, set the compass to a separation distance of 4 cm.

This will give a circle with a diameter of 8 cm.

To draw a circle with a diameter of 12 cm, set the compass to a separation distance equal to the radius of the circle.

As $d = 2 \times r$, the radius will be 6 cm.

👥 Draw this circle pattern.

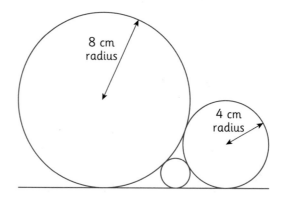

8 cm radius

4 cm radius

Guided practice

Draw a circle with a diameter of 16 cm.

To draw a circle with a diameter of 16 cm, I need to set a compass to a separation distance of 8 cm. This is equal to the radius of the circle.

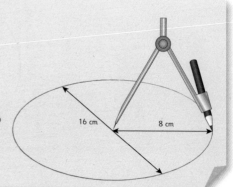

Lesson 4: **Rotational symmetry**

Key words
- **rotational symmetry**
- **order**

- Identify rotational symmetry in familiar shapes, patterns or images

Let's learn

An equilateral triangle rotated around its centre will appear as it did before the rotation three times every full rotation.

We say that an equilateral triangle has rotational symmetry of order 3.

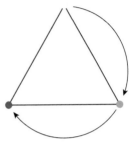

What is the order of rotational symmetry of each of these shapes?

a

b

Geometry and Measure

Guided practice

What is the order of rotational symmetry of this shape?

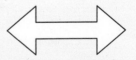

I rotate the shape around its centre a full turn and count the number of times it appears as it did before the rotation.

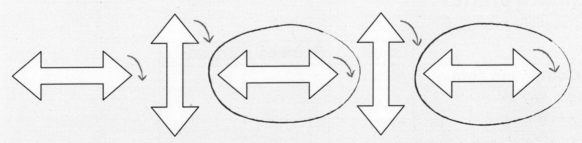

The shape has rotational symmetry of order 2.

Lesson 1: **Identifying and describing compound 3D shapes**

• Identify and describe compound 3D shapes

Let's learn

A **compound 3D shape** is a structure built from two or more component 3D shapes.

Here are some examples.

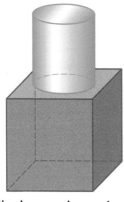

A cylinder and a cube

Two cuboids

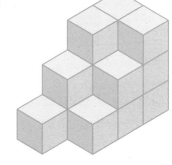

👥 Find the smallest number of unit cubes needed to turn this shape into a cuboid.

Guided practice

Complete the table.

	Faces	Vertices	Edges
Compound shape	8	12	18

Lesson 2: **Sketching compound 3D shapes**

- Sketch compound 3D shapes

Let's learn

When sketching a compound 3D shape on plain paper, think about the 2D faces that form the shapes.

Sides that are not visible can be shown by drawing dashed lines.

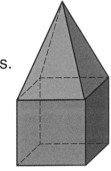

Sketching on triangular dot paper makes it easier to see where the edges of a shape are positioned.

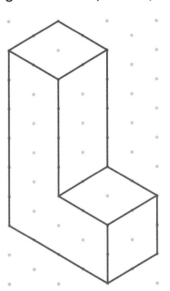
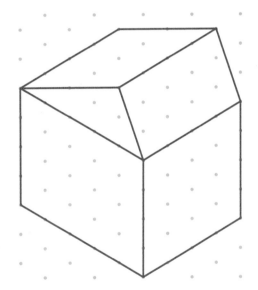

Sketch a T-shaped compound 3D shape.

The shape is built from two cuboids.

One cuboid is balanced around its midpoint on top of a second cuboid arranged at right angles with the first.

Guided practice

Sketch a compound shape made from three cubes connected together.

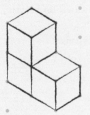

Geometry and Measure

Lesson 3: **Identifying nets**

Key word
• net

• Identify the nets for different 3D shapes

Geometry and Measure

Let's learn

A **net** is what a 3D shape would look like if it was opened out flat.

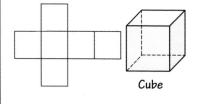
Cube

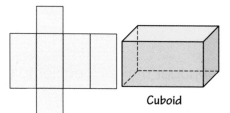
Cuboid

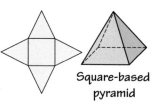
Square-based pyramid

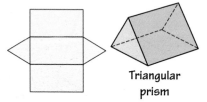
Triangular prism

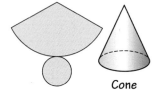
Cone

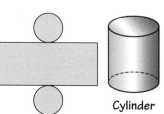
Cylinder

👥 You will need a cereal packet and some sticky tape.

6 Fold back the faces of a cereal packet and flatten out the card to form a net of a cuboid.

Cut the card to separate the individual faces of the shape.

Try to rearrange and reattach the faces with sticky tape so that they form another net.

Investigate if there are other nets possible for a cuboid.

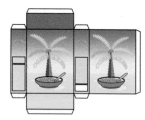

Guided practice

Draw a ring around the net for a triangular prism.

Lesson 4: **Sketching nets**

Key word
• net

• Sketch nets for different 3D shapes

Let's learn

To sketch a net for a 3D shape, it is important to look at the shape from different viewpoints, for example from the front, the side and below.

This will help you to work out the position and shape of each face and how the faces connect.

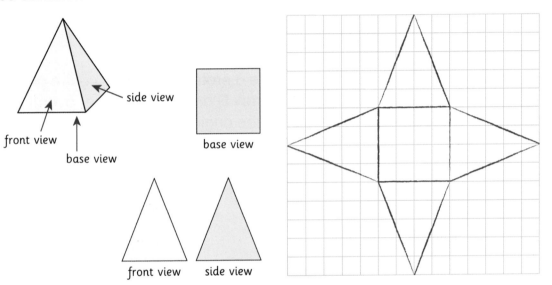

front view   base view   side view   base view   front view   side view

Geometry and Measure

What shape is this sweet tin?

Sketch the net for this shape.

Guided practice

Sketch the net for this shape.

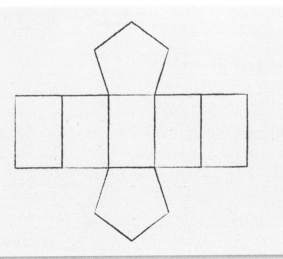

Lesson 1: **Measuring angles**

- Classify, estimate and use a protractor to measure angles

Let's learn

Key words
- angle
- acute
- right
- obtuse
- straight
- reflex
- protractor
- benchmark angles

Full-circle protractor

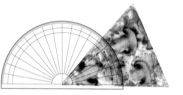

Half-circle protractor

Measuring degrees

Use a protractor with its centre point on vertex **B** and 0° (base line) over the arm of the angle (**AC**). Measure the angle between the arms.

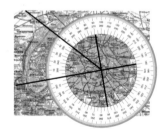

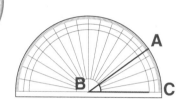

👥 Use a protractor to measure angles **a** to **f**.

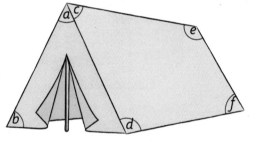

Guided practice

Use a protractor to measure angle B.

I estimate first. Angle B looks to be less than 45°.

I measure angle B. The angle is 35°.

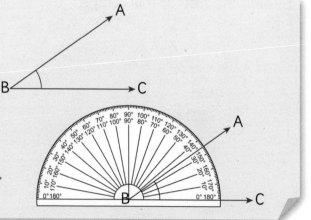

Lesson 2: **Drawing angles**

• Use a protractor to draw angles

Let's learn

There are four steps to drawing an angle.

Example: Draw an angle that measures 127°.

Step 1: Use a ruler to draw a straight line. This is one arm of the angle.

Step 2: Mark a dot at one end of the line – the vertex.
Centre the protractor over the dot and the baseline over the arm of the angle.

Step 3: Mark a dot at 127° on the scale. Remove the protractor.

Step 4: Join the dot to the vertex with a ruler to form the angle 127°.

Use a protractor to measure the size of angle A.

Now use the protractor and a ruler to draw angle A.

Discuss with your partner the size of each angle if you divide angle A equally into four smaller angles.

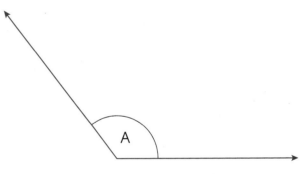

Guided practice

Use a protractor to draw an angle that measures 164°.

I estimate first.
Since 164° is
nearly 180° I expect
the angle to be
nearly a large obtuse
angle close to 180°.

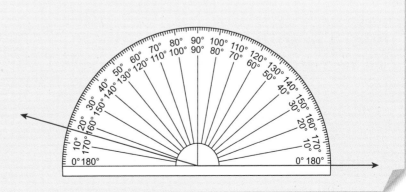

Geometry and Measure

Lesson 3: **Calculating missing angles in a triangle (1)**

- Calculate missing angles in a triangle

Let's learn

How many degrees are in a triangle?

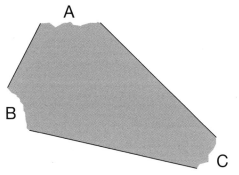

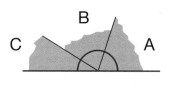

Rule: The sum of the interior angles of any triangle equals 180°.

Find the missing angle, A.

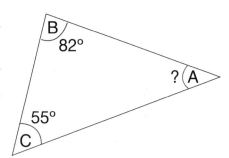

$$A + B + C = 180°$$
$$A = 180° - (B + C)$$
$$\quad = 180° - (82° + 55°)$$
$$\quad = 180° - 137°$$
$$\quad = 43°$$

Angle A is 43°.

👥 Use what you know about angles on a straight line and angles in a triangle to find the missing angle.

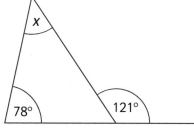

Guided practice

Find the missing angle x.

$68° + 47° + x = 180°$

$115° + x = 180°$

$x = 180° - 115°$

$x = 65°$

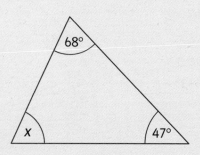

Lesson 4: **Calculating missing angles in a triangle (2)**

• Calculate missing angles in a triangle

Let's learn

Isosceles   Equilateral   Scalene

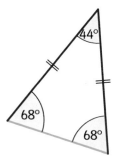

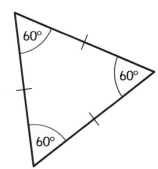

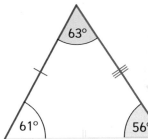

Two of the angles in an isosceles triangle are equal.

All angles are equal in an equilateral triangle. Each angle is 60°.

A scalene triangle has all unequal angles.

Geometry and Measure

👥 Work out the size of angle x.

3

4

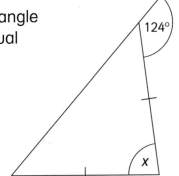

Guided practice
Find the missing angle e.

I know the triangle is isosceles since it has two equal sides AB and AC.

Angles B and C are base angles and are therefore equal.

The sum of the angles of a triangle is 180°,
so I know that A + B + C = 180°.

66° + B + C = 180°

B + C = 180° – 66°

I estimate by rounding: 180 – 70 = 110.

B + C = 114

As B = C, both angles must be 57°.

Angle e is 57°.

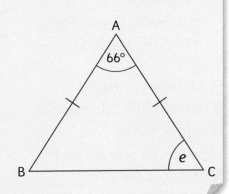

Lesson 1: **Converting time intervals (1)**

> ## Key words
> * **second**
> * **minute**
> * **hour**

* Understand the relationship between units of time, and convert between them, including times expressed as a fraction or decimal

Let's learn

How many minutes and hours is 6·75 hours?

$0.75 \, h = \dfrac{3}{4}$ of a hour

$ = \dfrac{3}{4}$ of 60 min

$ = 45$ min

$6.75 \, h = 6 \, h \; 45 \, min$

0·25 or $\dfrac{1}{4}$ hour

15 minutes

0·5 or $\dfrac{1}{2}$ hour

30 minutes

0·75 or $\dfrac{3}{4}$ hour

45 minutes

1 hour

60 minutes

It's easy to convert minutes to seconds.

For example, 3·5 minutes is 3 minutes and 50 seconds.

Is Mandisa correct? If not, explain the mistake she has made.

How would you correct Mandisa's thinking?

Guided practice

Convert 7·25 minutes to minutes and seconds.

$7.25 = 7 + 0.25$

1 minute = 60 seconds

$0.25 = \dfrac{1}{4}$

$0.25 \, min = \dfrac{1}{4} \times 60 \, s = 15 \, s$

$7.25 \, min = 7 \, min \; 15 \, s$

Lesson 2: **Converting time intervals (2)**

* Understand the relationship between units of time, and convert between them, including times expressed as a fraction or decimal

one tenth of an hour
$\frac{1}{10}$ = 6 minutes

Let's learn

How many minutes and hours is 6·7 hours?

$$0.7 \text{ h} = \frac{7}{10} \text{ of an hour}$$

$$\frac{1}{10} \text{ of an hour} = 60 \text{ minutes} \div 10$$
$$= 6 \text{ min}$$

$$\frac{7}{10} \text{ of an hour} = 7 \times 6 \text{ min}$$
$$= 42 \text{ min}$$
$$6.7 \text{ h} = 6 \text{ h } 42 \text{ min}$$

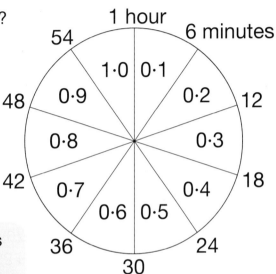

It's easy to convert hours to minutes.

For example, 8·2 minutes is 8 minutes and 20 seconds.

Is Danny correct? If not, explain the mistake he has made.

How would you correct Danny's thinking?

Guided practice

Convert 4·9 minutes to minutes and seconds.

4·9 = 4 + 0·9

1 minute is 60 seconds

$0.9 = \frac{9}{10}$

$\frac{1}{10}$ of 60 seconds = 60 ÷ 10 = 6 s

$\frac{9}{10}$ of 60 seconds = 9 × 6 s = 54 s

4·9 mins = 4 mins 54 s

Geometry and Measure

87

Lesson 3: **Capacity and volume (1)**

- Understand the difference between volume and capacity

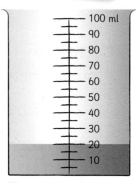

Key words
- volume
- capacity
- millilitre
- litre

Geometry and Measure

Let's learn

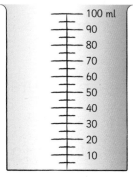

The capacity of the beaker is 100 ml

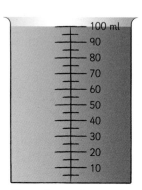

The volume of water in the beaker is 100 ml

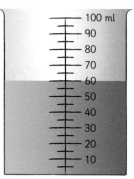

The volume of water in the beaker is 60 ml

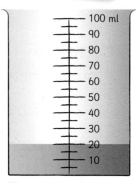

The volume of water in the beaker is 20 ml

6 Eleanor is holding up a bottle of fizzy orange drink labelled 250 ml.

Which of the following statements are **sometimes true**, **always true** or **never true**? Why?

a The capacity of the bottle is 250 ml.

b The volume of orange in the bottle is 150 ml.

c The capacity of orange in the bottle is 250 ml.

d The volume of the bottle is 250 ml.

e The maximum volume of orange that the bottle can hold is 250 ml.

Guided practice

Write the capacity of the jug and the volume of water it holds.

Capacity: [100] ml

Volume: [45] ml

Lesson 4: **Capacity and volume (2)**

* Understand the difference between volume and capacity

Let's learn

In order to solve problems that involve capacity, it is often necessary to convert between litres and millilitres.

Buckets A and B have capacities $5l\,262\,ml$ and $3{\cdot}8\,l$.
What is the total capacity of the buckets?

$5l\,262\,ml + 3{\cdot}8\,l = 5{\cdot}262\,l + 3{\cdot}800\,l$
$\qquad\qquad\qquad = 9{\cdot}062\,l$

or

$5l\,262\,ml + 3{\cdot}8\,l = 5262\,ml + 3800\,ml$
$\qquad\qquad\qquad = 9062\,ml$
$\qquad\qquad\qquad = 9{\cdot}062\,l$

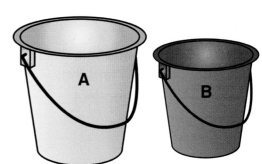

A 5-litre watering can and a 10-litre bucket both contain water. The volume of water in each is shown on the container.

What is the total volume of water in the two containers?

Give your answer in litres and millilitres.

$2l\,455\,ml$ $6{\cdot}25\,l$

Guided practice

Jugs A and B have capacities of $2l\,445\,ml$ and $1{\cdot}78\,l$.
What is the total capacity of the jugs?

To add the capacities, convert them to the same units.

$2l\,445\,ml = 2{\cdot}445\,l$

Total capacity of the two jugs = $2{\cdot}445\,l + 1{\cdot}78\,l$.
The total capacity of the jugs is $4{\cdot}225\,l$.

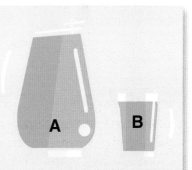

A   B

```
  2 . 4 4 5
+ 1 . 7 8 0
  4 . 2 2 5
    1   1
```

Geometry and Measure

Lesson 1: Calculating the area of a triangle (1)

Key words
- area
- right-angled triangle

- Prove that the area of a right-angled triangle is half the area of its related rectangle

Let's learn

Cut along the diagonal to form two right-angled triangles.

By arranging one triangle over the other, you find that the triangles are identical

This tells you that the area of a triangle is equal to half the area of a rectangle around it.

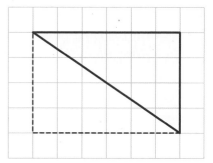

Without counting squares, what is the area of the yellow triangle?

Guided practice

Draw two rectangles on squared paper with different dimensions that both have an area of 12 square units.

Mark each rectangle to create a triangle with an area of 6 square units.

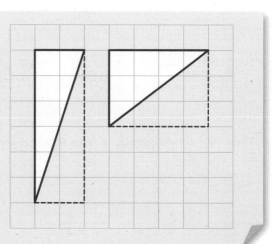

Lesson 2: **Calculating the area of a triangle (2)**

Key words
- area
- right-angled triangle

- Use the area of a related rectangle to find the area of a right-angled triangle

Let's learn

The area of this rectangle is 8 cm².

4 cm

2 cm

So the area of each right-angled angle triangle is 4 cm².

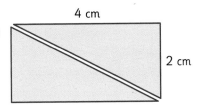

4 cm

2 cm

6 cm

👤 What is the area of each triangle?

5 cm

Geometry and Measure

Guided practice

What is the area of each triangle?

I know that the area of each triangle is half the area of the rectangle.

Area of rectangle = 10 cm × 8 cm = 80 cm²

Area of triangle = $\frac{1}{2}$ × 80 cm² = 40 cm²

The area of each triangle is 40 cm².

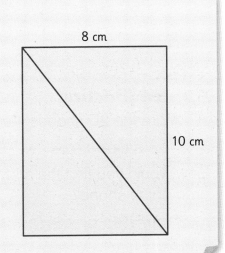

8 cm

10 cm

Lesson 3: **Surface area (1)**

Key words
- area
- surface area

- Understand the relationship between area of 2D shapes and surface area of 3D shapes

Geometry and Measure

Let's learn

Surface area is the sum of all the areas of all the shapes that cover the surface of a 3D shape.

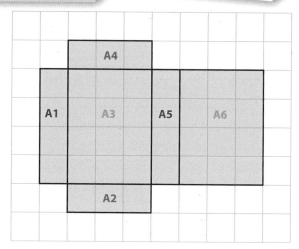

Surface area of cuboid = A1 + A2 + A3 + A4 + A5 + A6

👥 Talk to your partner about the different 2D shapes that cover the surface of this 3D shapes. Which faces of the 3D shape are the same?

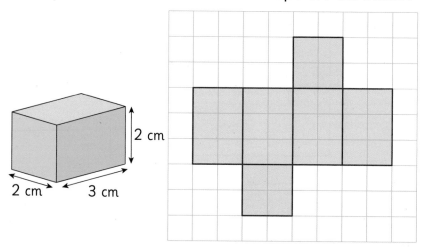

Guided practice

What are the 2D shapes that cover the surface of this 3D shape?

6 identical squares

How would you find the surface area of the shape?

Find the area of one face and multiply it by 6.

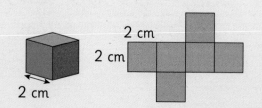

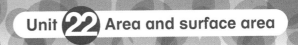
Lesson 4: **Surface area (2)**

- Understand the relationship between area of 2D shapes and surface area of 3D shapes

Let's learn

The surface area of the puzzle cube can be found by multiplying the area of one face by 6.

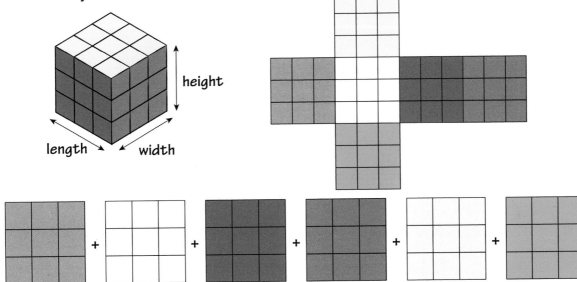

height

length   width

👥 Talk to your partner about the different 2D shapes that cover the surface of this 3D shapes. How might you work out the surface area of the shape?

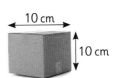

10 cm

10 cm

Geometry and Measure

Guided practice

What are the different 2D shapes that cover the surface of each 3D shape? How many of each 2D shape are there?

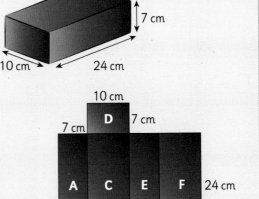

7 cm

10 cm   24 cm

There are six rectangles.

Which shapes have the same area?

Rectangles A and E have the same area.

Rectangles B and D have the same area.

Rectangles C and F have the same area.

Lesson 1: **Reading and plotting coordinates (1)**

- Read coordinates in all four quadrants

Let's learn

A set of coordinates can be written as positive or negative integers, decimals or fractions.

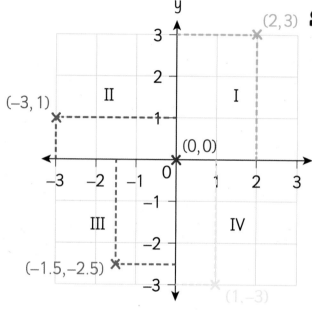

A children's treasure hunt is being planned on a map drawn on a coordinate grid.

A compass on the map shows that the *y*-axis points directly north.

Two 'treasures', A and B, are located at coordinates (4, 6) and (−2, −3).

The next two 'treasures', C and D, must be located at points so that:

- treasure C is south of treasure A and in Quadrant IV
- treasure D is north of treasure B and in Quadrant II.

Provide possible coordinates for the locations of treasures C and D.

Geometry and Measure

Guided practice

Identify the coordinates of points E, F and G.

E (−6, 7)

F (−3, −5)

G (2·5, −2)

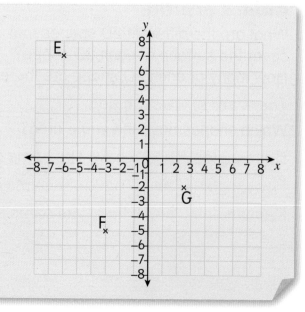

Lesson 2: **Reading and plotting coordinates (2)**

- Plot coordinates in all four quadrants

Let's learn

1. Always begin at the Origin (0, 0).

2. The x tells you to move left or right. A negative number tells you to move left and a positive number tells you to move right.

3. The y tells you to move up or down. A negative number tells you to move down and a positive number tells you to move up.

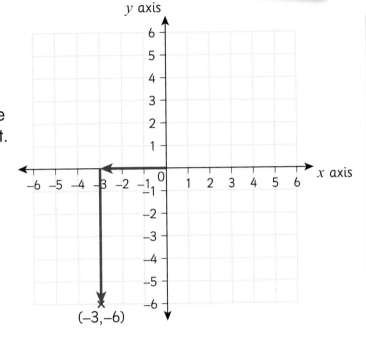

(–3,–6)

Without plotting the following points, sort them into four groups according to the quadrant they lie in.

A (–3, 4)   B (9$\frac{1}{4}$, 1)   C (5, –7)   D (–8, –1·5)

E (2, –5)   F (–9·75, –9)   G (1, 3)   H (–9, –6)

Geometry and Measure

Guided practice

Plot the following points:

J (–8, –3)   K (–7, 5)   L (4, –9·5)

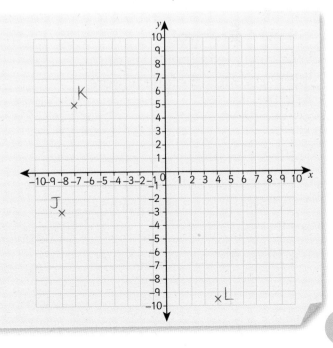

Lesson 3: **Plotting lines and shapes across all four quadrants (1)**

- Plot points to form shapes in all four quadrants

Geometry and Measure

Let's learn

Properties of 2D shapes can help locate the missing vertex of a shape on a coordinate plane.

The coordinates for three vertices of a rectangle are A (6, –2), B (6, –7), C (–4, –7).

The missing vertex D will have the same *x*-coordinate as vertex C (–4) and the same *y*-coordinate as vertex A (–2).

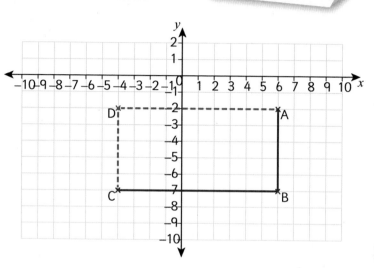

So, vertex D has the coordinates (–4, –2).

An isosceles triangle ABC has a base AB with vertices at points A (–4, –8) and B (8, –8).

If the triangle has a height of 11 units and vertex C has a positive *y*-coordinate, what is the position of vertex C?

Guided practice

What are the coordinates of the missing vertex D of the square ABCD?

The square has sides of 6 units.

To find the missing vertex, I move 6 squares down from vertex C (or 6 squares right from vertex A)

The coordinates of vertex D are (2, 3).

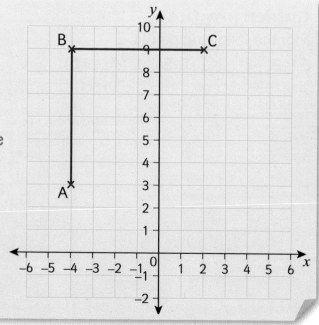

Lesson 4: Plotting lines and shapes across all four quadrants (2)

- Plot points to form lines in all four quadrants

Let's learn

A line is the shortest distance between two points.

Two points that lie on line AB are C (1, –2) and D (2·5, 0).

Points F and G lie on a line: F (–6, 4), G (6, –2).

Point H is on the line. Its *x*-coordinate is 2. What is its *y*-coordinate?

Point K is on the line. Its *y*-coordinate is 0. What is its *x*-coordinate?

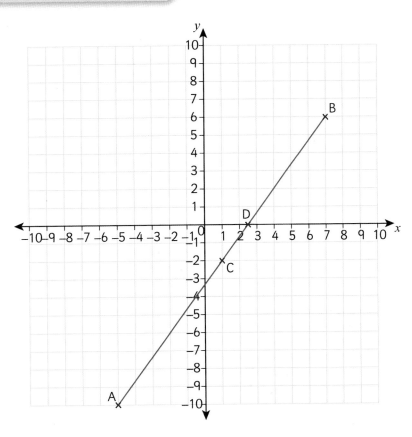

Geometry and Measure

Guided practice

Points M and N lie on a line: M (–2, –6), N (1, 6).

Plot and join the points with a straight line and write the coordinates of two other points that lie on the same line.

S (–1, –2)

T (0, 2)

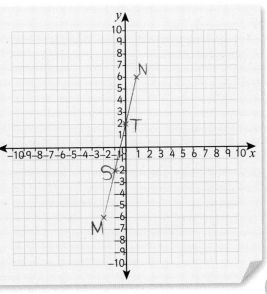

Lesson 1: **Translating 2D shapes on coordinate grids**

- Translate shapes across all four quadrants on a coordinate grid

Geometry and Measure

Let's learn

Translate the triangle ABC 8 units right and 7 units up.

You usually label the image of the vertex of a shape using the prime symbol (').

For example, the image of vertex A should be labelled A'.

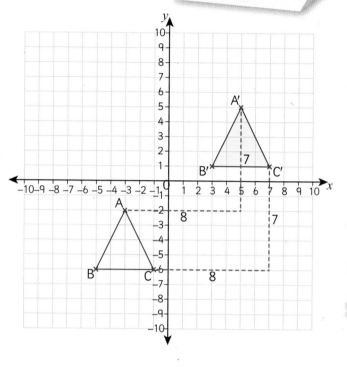

👥 Describe the translation that moves the vertices of a square ABCD from position 1 to position 2:

Position 1: A (−3, −3), B (2, −3), C (2, −8), D (−3, −8)

Position 2: A' (−8, 9), B' (−3, 9), C' (−3, 4), D' (−8, 4)

Guided practice

Translate rectangle ABCD 6 units left and 10 units down.

Write the coordinates of the corresponding vertices in the image A', B', C' and D'.

A' (−4, −3)   B' (−4, −5)

C' (1, −5)   D' (1, −3)

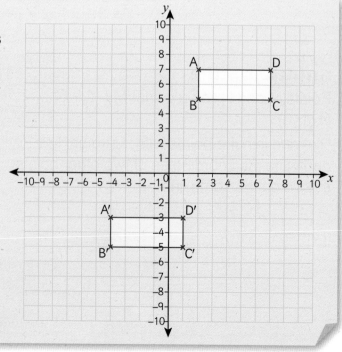

Lesson 2: **Reflecting 2D shapes in a mirror line (1)**

• Reflect shapes in horizontal and vertical mirror lines

Let's learn

To reflect a shape, reflect the vertices and join them to form the image.

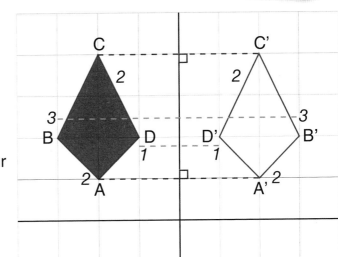

1 Measure the perpendicular distance from the vertex to the mirror line.

2 Measure the same perpendicular distance on the other side and plot the vertex.

3 Connect the vertices to reflect the original object.

Frankie's teacher asked him to reflect a triangle in a horizontal mirror line.

Has Frankie completed the task correctly?

If not, what mistakes has he made?

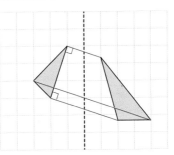

Geometry and Measure

Guided practice
Reflect rectangle ABCD in the mirror line.

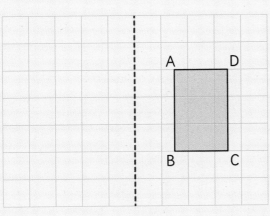

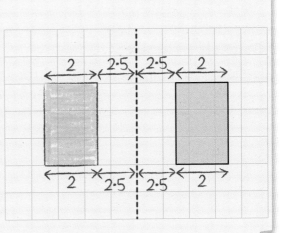

Lesson 3: **Reflecting 2D shapes in a mirror line (2)**

Key words
* reflect
* vertex (vertices)
* perpendicular distance
* diagonal

* Reflect shapes in diagonal mirror lines

Geometry and Measure

Let's learn

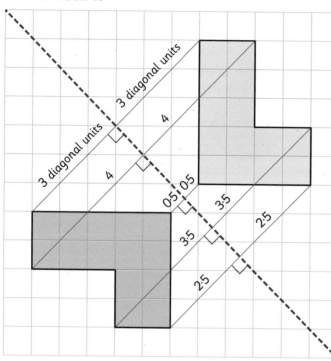

Alice's strategy

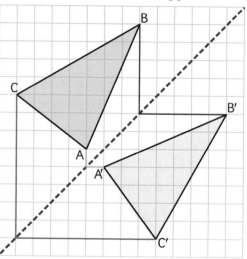

4 Alice has found a different strategy for reflecting a shape across a diagonal mirror line.

Critique this method. How could it be improved?

Guided practice

Reflect the rectangle in the mirror line.

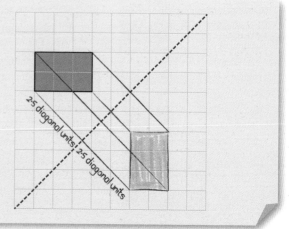

Lesson 4: **Rotating shapes 90° around a vertex**

- Rotate shapes 90° around a vertex (clockwise or anticlockwise)

Key words
- rotate
- centre or point of rotation
- vertex (verices)
- clockwise
- anticlockwise

Let's learn

A rectangle rotated around a vertex through four turns of 90°.

You are given shape A.

How would you use the shape to create the pattern shown under shape A?

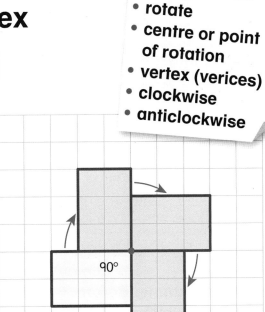

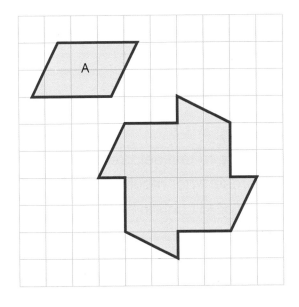

Geometry and Measure

Guided practice

Draw the image of the square under a rotation of 90° anticlockwise about vertex A.

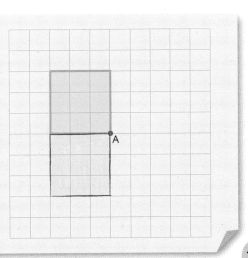

Lesson 1: **Venn and Carroll diagrams**

Key words
• statistical question
• prediction
• data
• Venn diagram
• Carroll diagram

• Record, organise, represent and interpret data in Venn and Carroll diagrams

Let's learn

I want to find out if there is a difference between the number of boys and girls who have earlier bedtimes (before 8.30 p.m.) and those who have later bedtimes.

I predict that girls are allowed to stay up later!

Use a Carroll diagram to show the data.

Two columns for 'earlier'/'not earlier' bedtimes and two rows for 'boys'/'not boys'.

	Bedtime: Before 8.30 p.m.	Bedtime: Not before 8.30 p.m.
Boy		
Not boy		

Maha thinks that more boys own bikes than girls.

Write a statistical question that she could investigate to test her prediction.

Draw a Carroll diagram that she could use to represent her results.

What information would she need to look for in order to confirm her prediction?

Guided practice

Liam wants to know if more men are likely to own red cars than people who are not men.

Liam predicts that more men are likely to choose red cars.

	Own a red car	Do not own a red car
Men	× × × × × × × × × × × × ×	× × × × × × × × × × ×
Not men	× × × × × × × × × × × ×	× × × × × × × × × × × ×

• There is an even split between 'not men' who own a red car and those who do not own a red car.

• For men, the split is 13:11, which is very close to an even split 12:12.

• The data disagrees with Liam's prediction. However, if a larger number of people were surveyed, it would give a more reliable result.

Lesson 2: **Tally charts, frequency tables and bar charts**

• Record, organise, represent and interpret data in tally charts, frequency tables and bar charts

Let's learn

The 30 learners in Class 6A took a Maths test. Their marks out of 20 were recorded.

Freya predicted that over 50% of learners would get over 50% of the answers correct (11 marks or more).

The results of the test were presented in a bar chart.

Number of learners who scored over 50% (11 marks or more) = 12 + 7 = 19.

$\frac{19}{30}$ learners scored over 50%, so Freya's prediction is correct.

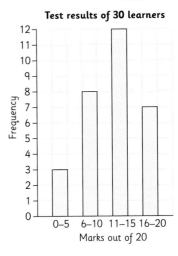

Test results of 30 learners

Find a book aimed at children of your age.

3 Choose one page at random.

Make a prediction about the frequency of words with different numbers of letters, for example, 'I think words with three, four and five letters will be the most frequent.'

Test your prediction. Draw a frequency table with numbers of letters organised into intervals. Go through the page, counting the number of letters in each word.

Draw a bar graph to represent the data. Does the data confirm your prediction? How do you know?

Guided practice

Billy works in a garden centre and investigates the growth of 800 plants. He predicts that at least 60% of the plants will grow to a height of more than 40 cm after two weeks.

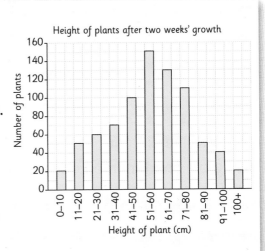

Height of plants after two weeks' growth

The graph shows the results after two weeks of growth. Is Billy's prediction correct? How do you know?

From the graph, I can see that the number of plants that grew to a height of 41 cm or more = 100 + 150 + 130 + 110 + 50 + 40 + 20 = 600.

% of plants that grew to this height $\frac{600}{800} = \frac{6}{8} = \frac{3}{4} = 75\%.$

Statistics and Probability

Lesson 3: **Waffle diagrams and pie charts**

> **Key words**
> * statistical question
> * waffle diagram
> * pie chart

* Record, organise, represent and interpret data in waffle diagrams and pie charts

Statistics and Probability

Let's learn

The table shows the results of a class survey that asked 10 learners the question:

How many brothers and sisters do you have?

The frequencies have been converted to percentages.

Number of brothers and sisters	Frequency	Percentage
0	2	$\frac{2}{10} \times 100\% = 20\%$
1	3	$\frac{3}{10} \times 100\% = 30\%$
2	3	$\frac{3}{10} \times 100\% = 30\%$
3	1	$\frac{1}{10} \times 100\% = 10\%$
> 3	1	$\frac{1}{10} \times 100\% = 10\%$

We can use the percentages to complete:

i a waffle diagram to show the results of the survey

ii a pie chart marked out in 10 equal divisions.

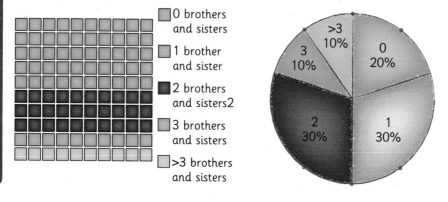

Work with a partner to complete your own survey of the approximate distance learners live away from the school.

Choose intervals, for example: 0–1 km, 1–2 km, 2–3 km, and so on. You may need to use the internet or a map to work this out.

Predict which distance will be the most frequent.

Ask 20 learners to say how far they live away from school and record the data in a tally chart. Find the frequency of each distance and then convert these values to percentages.

Present the data in a waffle diagram. Use the data to construct a pie chart.

What conclusions can you draw from the charts? Do they confirm your prediction?

Guided practice

The table shows the data collected when 10 learners were asked how many siblings (brothers and sisters) they have.

Complete the table then represent the data in a pie chart and use it to draw conclusions about the results of the survey.

I complete a pie chart marked out in 10 equal divisions using the information.

I can see that most learners have 2 or more siblings.

Very few learners have no siblings or more than 3 siblings.

Number of siblings	Frequency	Percentage
0	1	$\frac{1}{10} \times 100\% = 10\%$
1	2	$\frac{2}{10} \times 100\% = 20\%$
2	4	$\frac{4}{10} \times 100\% = 40\%$
3	2	$\frac{2}{10} \times 100\% = 20\%$
> 3	1	$\frac{1}{10} \times 100\% = 10\%$

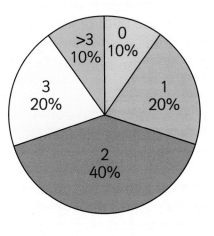

Statistics and Probability

Lesson 4: **Mode, median, mean and range**

Key words
- mode
- median
- mean
- range

- Find and interpret the mode, median, mean and range of a data set

Let's learn

Mode, median and mean are types of average.

Data set:

28 cm	26 cm	24 cm	27 cm	24 cm
26 cm	24 cm	25 cm	25 cm	26 cm

Average	Description	Calculation
Mode	The most frequent value or values.	24 and 26
Median	Order the data from lowest to highest: 24, 24, 24, 25, 25, 26, 26, 26, 27, 28 The median is the middle value or the sum of two middle values divided by 2.	$(25 + 26) \div 2$ $= 51 \div 2$ $= 25 \cdot 5$
Mean	The sum of the values divided by the number of values.	$24 + 24 + 24 + 25 + 25 + 26 + 26 + 26 + 27 + 28 = 255$ $255 \div 10 = 25 \cdot 5$
Range	The highest value subtract the lowest value.	$28 - 24 = 4$

👥 This list shows the amounts spent by 7
4 families when buying a car:

$900, $1100, $16 400, $1000, $14 600
$1000, $700

a Calculate the mode, median, mean and range of these amounts.

b Suggest which of the three averages is the most useful indicator of how much a typical family spends on a car.

Guided practice

Aisha is thinking of four numbers.

> The mode of the four numbers is 9.
> The median is 8.5.
> The mean is 8.
> What are the numbers?

I can identify the numbers by using what I know about mode, median and mean.

Mode: If 9 is the mode, then two of Aisha's numbers will be 9. We cannot have three 9's, because this would affect the median. The numbers are therefore 9, 9, x, x

Median: If the median is 8.5, Aisha's third number is 8, because 8.5 is exactly half-way between 8 and 9. The numbers are now x, 8, 9, 9

Mean: If the mean is 8, then the total of Aisha's numbers is 32, because $8 = 32 \div 4$. At the moment, Aisha's numbers total $8 + 9 + 9 = 26$. This means that the last remaining number is 6, making Aisha's numbers 6, 8, 9, 9

Lesson 1: **Frequency diagrams**

- Plan and conduct an investigation for a set of related statistical questions
- Predict the answer to a statistical question
- Represent data in a frequency diagram

Key words
- **statistical question**
- **frequency diagram**

Let's learn

I want to know the range of arm spans (the distance from fingertip to fingertip of a person's outstretched arms) in my class.

I predict that the most common arm span is around 135 cm.

Use a **frequency diagram** to show the data.

The graph shows that the most common arm span is in the range 130 to 140 cm.

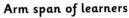

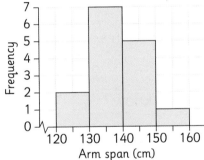

Arm span of learners

Arm span (cm)	Frequency
120–130	2
130–140	7
140–150	5
150–160	1

Florence thinks that most daffodils in her flowerbed are between 50 and 60 cm in height.

Write a statistical question that she could investigate to test her prediction.

Draw a data collection table and a frequency diagram that she could use to represent her results.

What information would she need to look for in order to confirm her prediction?

Guided practice

Statistical question: Rachel wants to know the typical bill for the tables at her restaurant today.

Rachel says, 'I predict it will be around $130.'

Bill ($)	Frequency
60–80	4
80–100	6
100–120	9
120–140	7
140–160	2

The results of a survey to answer the question are given in the table.

Is Rachel's prediction correct?

- *I can see that most people have paid between $100 and $120 for their meals.*
- *Rachel's prediction isn't correct but it is close.*

Statistics and Probability

107

Lesson 2: **Line graphs**

- Plan and conduct an investigation for a set of related statistical questions
- Predict the answer to a statistical question
- Represent data in a line graph

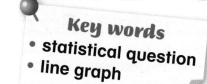

Key words
- statistical question
- line graph

Let's learn

A **line graph** displays data that changes over time, such as the growth of a plant or the distance a cyclist travels.

The data can take on any value, for example: 7, 7·4, 7·43.

This graph shows the height of a bean plant over several days.

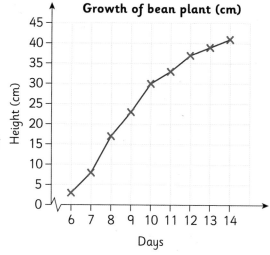

Growth of bean plant (cm)

3 The graph in the Let's learn section shows the growth of a bean plant.

  a Between which days did the bean plant show the greatest rate of growth? How do you know?

  b What do you think will happen to the rate of growth after 14 days?

Guided practice

Zoe the zookeeper throws some fish into a pool for penguins to eat.

She counts the number of penguins feeding in the pool over a 10-minute period.

Time (min)	0	2	4	6	8	10
Number of penguins	47	43	35	24	11	1

Draw a line graph using the data. What does this information tell you about how penguins feed?

The penguins feed intensely to begin with, then the numbers in the pool drop off as the amount of fish left to feed on decreases.

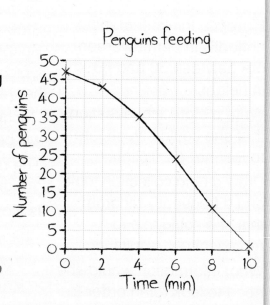

Penguins feeding

Statistics and Probability

Lesson 3: **Scatter graphs**

- Plan and conduct an investigation for a set of related statistical questions
- Predict the answer to a statistical question
- Represent data in a scatter graph
- Know how to draw a line of best fit

Key words
- statistical question
- scatter graph
- line of best fit

Let's learn

A **scatter graph** displays two sets of data.

You can use the graph to see if there is a relationship between the data sets.

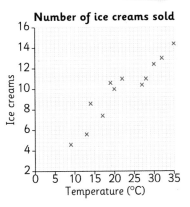
Number of ice creams sold

Millie drew a line of best fit through the plotted points for the ice cream data in the Let's learn section.

What questions could you use the graph to answer? Write three and answer each one.

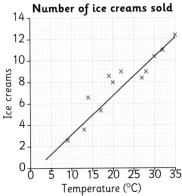
Number of ice creams sold

Guided practice

The table shows the number of people who visited a beach café over a 10-day period in the summer.

Hours of sunshine	4	10	2	8	5	7	1	9	6	3
Number of visitors	42	96	23	77	48	70	11	92	64	29

Plot the points to create a scatter graph and draw a line of best fit. What does this information tell you about the relationship between the number of hours of sunshine and the number of visitors to the café?

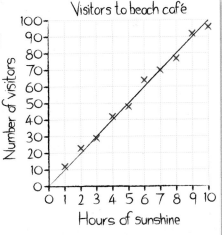
Visitors to beach café

The relationship between the two measurements is directly proportional. This means the more sunshine there is, the more people visit the café.

Statistics and Probability

109

Lesson 4: **Dot plots**

* Plan and conduct an investigation for a set of related statistical questions
* Predict the answer to a statistical question
* Represent data in a dot plot

Let's learn

I want to graph the data I've collected to answer the statistical question: How many pets do learners have at home?

I want a graph that is easy to read and shows peaks and clusters, and any gaps.

A **dot plot** is one of the best graphs to show the spread of data and where it might cluster together.

Statistics and Probability

Number of pets owned by learners

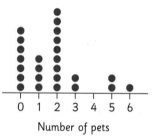

Number of pets

 Write three conclusions that you can draw from the dot plot in the Let's learn section.

Use these words in your answers: 'peak', 'cluster' and 'gap'.

How would you explain these features of the graph?

What questions could you use the graph to answer? Write three and answer each one.

Guided practice

The learners in Badger Class were asked to read a book and record how many pages they read. The dot plot shows the results.

Pages of a book read

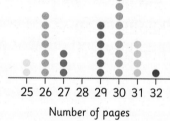

25 26 27 28 29 30 31 32
Number of pages

a Which number of pages read has the highest frequency? 30 pages

The lowest frequency? 32 pages

b What is the range of the data? 7

c Does the data show any peaks, clusters or gaps? If so, where?

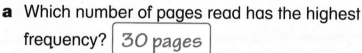

There are peaks at frequencies of 26 pages and 30 pages, and two clusters, one from 25 and 27 pages and another from 29 and 31 pages. There is also a gap at a frequency of 28 pages.

Lesson 1: **Describing and comparing outcomes**

Key words
- **probability**
- **proportion**
- **percentage**
- **possible outcomes**

- Use the language of proportion and probability to describe and compare the probability of different outcomes

Let's learn

Outcome	Probability
spinning orange	1 in 4 (25%)
spinning blue or red	2 in 4 or 1 in 2 (50%)
spinning blue, yellow or red	3 in 4 (75%)

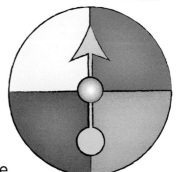

Give your answers as a proportion and/or a percentage.

What is the probability of spinning a:

a 2?

b number greater than 4?

c number less than 3?

d number that is not 7 or 8?

Which is greater, the probability of spinning a number greater than 3 or the probability of spinning an even number? How do you know?

Guided practice

Give your answers as a proportion and a percentage.

What is the probability of picking a:

a green bead from the bag?

    | 3 | in | 10 | or | 30 |%

b blue bead from the bag?

    | 7 | in | 10 | or | 70 |%

Statistics and Probability

Lesson 2: **Independent and mutually exclusive events**

Key words
- event
- mutually exclusive
- independent

- Identify when two events are mutually exclusive or independent

Let's learn

Mutually exclusive events	Independent events
Events that **cannot** happen at the same time.	Events that **can** happen at the same time.
The outcome of one event **does** influence the outcome of the other.	The outcome of one event **does not** influence the outcome of the other.
The outcome of rolling a 1 to 6 dice.	The outcome of rolling a 1 to 6 dice and flipping a coin.
2 4 6 — even 1 3 5 — odd	odd or even heads or tails
The outcome cannot be both even and odd.	The outcome can be odd or even AND heads or tails.

Statistics and Probability

Which of these pairs of events are mutually exclusive?

a Winning a chess game and drawing the same game

b Wearing one green sock and one red sock

c Eating an ice cream in the morning and a cake in the afternoon

d Being on time and being late for school

Guided practice
Are the following events independent? Explain your answer.
Two 1 to 6 dice are rolled.
Dice A lands on a 4; dice B lands on a 4.

I think they are independent.

What number is rolled on one dice does not affect the number rolled on the other dice.

They can both be 4!

Lesson 3: **Event probability and the number of trials (1)**

Key words
- outcome
- trial
- probability experiment

- Predict and describe the frequency of outcomes using the language of probability
- Perform probability experiments using small and large numbers of trials

Let's learn

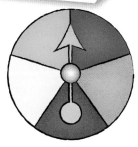

Spinning a colour spinner such as the one shown here is an example of a **probability experiment**.

As each of the five colour areas are the same size, you should expect the probability of spinning each colour to be 1 in 5 or 20%.

For example, the probability of spinning 'green' is 1 in 5.

This means that if you spin 20 times, you should expect to spin 'green' 4 times (20 ÷ 5 = 4).

If you spin 40 times, you should expect to spin 'green' 8 times (40 ÷ 5 = 8).

Jamie spins a letter spinner 80 times and records the results.

a How many times should he expect to spin the letter A? How do you know?

b How many times should he expect to spin the letter:

   i B?   **ii** C?

Guided practice

Jignesh rolls a 1 to 6 dice.

He rolls the dice 60 times and records the number each time.

a How many rolls would you expect Jignesh to record a '3'?

As there are 6 possible numbers that can be rolled, I know the chance of rolling any number is 1 in 6. Therefore, the chance of rolling a '3' is also 1 in 6. That means out of 60 rolls, I would expect to roll a '3' 10 times (60 ÷ 6 = 10).

b How many rolls would you expect Jignesh to record an even number?

There are 3 even numbers on the dice: 2, 4 and 6. That means there are 3 chances out of 6 to score an even number. $\frac{3}{6} = \frac{1}{2}$

That means out of 60 rolls, I would expect to roll an even number 30 times (60 ÷ 2 = 30).

Statistics and Probability

113

Lesson 4: **Event probability and the number of trials (2)**

* Predict and describe the frequency of outcomes using the language of probability
* Perform probability experiments using small and large numbers of trials

Let's learn

Tom flips a coin 50 times and records the outcomes.

He predicts that the coin will land on 'heads' 50% of flips and 'tails', 50% of flips.

These are the **predicted probabilities** – what he expects to happen.

However, Tom's record shows the **experimental probability**, what actually happens.

The probabilities $\frac{28}{50}$ and $\frac{22}{50}$ are close to 50% ($\frac{25}{50}$) but not exact.

Coin flip (outcome)	Frequency
heads	28
tails	22
total	50

If Tom increases the number of trials (flips), he should find that the experimental probability generally becomes closer to the predicted probability.

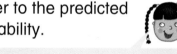

4
8

The probability of rolling a '2' on a 1 to 6 dice is 1 in 6. Over 6 rolls of a dice, I expect to roll '2' one time. Over 60 rolls of a dice, I expect to roll '2' ten times.

Plan and conduct an experiment to test Sophie's prediction.

Is the predicted probability of rolling a '2' the same as the experimental probability? How do you know?

If the two probabilities are different, then explain why this might be.

114 What could you do to bring the probabilities closer together?

Guided practice

If you spin a 1 to 4 spinner 100 times, what is the predicted probability of spinning an even number?

As there are four possible outcomes (1, 2, 3, 4), I know the predicted probability of spinning an even number (2 or 4) is 2 outcomes out of 4.

I would expect the frequency of spin of an even number to be 50 times out of 100 spins.

Perform a chance experiment to prove your predicted probability. What do the experimental results show?

The experiment shows the actual frequency to be 55 times out of 100 spins.

Spinner number (outcome)	Frequency
1	22
2	27
3	23
4	28

The Thinking and Working Mathematically Star

Think about an idea or solution and decide what's good and bad about it

- ◄ … is a good idea because …
- ◄ … is not a good idea because …
- ◄ … is right because …
- ◄ … is not right because …

Find an example and see if it fits a rule

- ◄ … is an example of …
- ◄ … is a good example because …
- ◄ … is not an example of …
- ◄ … is not an example because …

Ask questions and form ideas

- ◄ … I think ….
- ◄ … is always …
- ◄ … is sometimes …
- ◄ … is never …
- ◄ What if …

1 **2** Specialising and Generalising

3 **4** Conjecturing and Convincing

7 **8** Critiquing and Improving

Recognise patterns and find more examples

- ◄ The pattern is …
- ◄ The rule is …
- ◄ …. happens because …
- ◄ … is another example of …
- ◄ … are examples of … because …

Prove a mathematical idea or solution to others

- ◄ … works because …
- ◄ … is right because …
- ◄ I can prove …
- ◄ I can convince you that … because …
- ◄ I know … because …
- ◄ I know this is not correct because …

5 **6** Characterising and Classifying

Identify and describe a mathematical object or idea

- ◄ I have found out that …
- ◄ I know that …
- ◄ I notice that …
- ◄ … are the same because …
- ◄ … are different because …
- ◄ … is a property of …

Improve a mathematical idea to come up with a better explanation or solution

- ◄ If …
- ◄ … could be improved …
- ◄ … would be better because …
- ◄ Instead of … it would be better …
- ◄ Changing this improves it because …

Sort and organise mathematical objects or ideas into groups

- ◄ I can sort …
- ◄ I can organise …
- ◄ I have grouped …

The Thinking and Working Mathematically star, © Cambridge International, 2018

Acknowledgements

Photo acknowledgements

Every effort has been made to trace copyright holders. Any omission will be rectified at the first opportunity.

p12 Lunaticm/Shutterstock; p13 Ajt/Shutterstock; p16 Mas Hasyim/Shutterstock; p18l Mido Semsem/Shutterstock; p18r MaraZe/Shutterstock; p26 Sasha_Petrov/Shutterstock; p27 Amenic181/Shutterstock; p28 Shevchuk/Shutterstock; p29 Kup/Shutterstock; p31 Ugurr/Shutterstock; p32 Chalermpon Poungpeth/Shutterstock; p33 Lightspring/Shutterstock; p38b Mut Hardman/Shutterstock; p39 Lorelyn Medina/Shutterstock; p41 Sergey Novikov/Shutterstock; p46 Lorelyn Medina/Shutterstock; p52 ANNA ZASIMOVA/Shutterstock; p56t Mr.Timoty/Shutterstock; p56b Vladimir Kramin/Shutterstock; p57c Vector things/Shutterstock; p57b Sergey Novikov/Shutterstock; p58 Odua Images/Shutterstock; p63 Strajinsky/Shutterstock; p64 Graham Friend/Shutterstock; p66 Yeamake/Shutterstock; p67 Bradley Blackburn/Shutterstock; p81b Robuart/Shutterstock; p82bl Katstudio/Shutterstock; p81bc Josefkubes/Shutterstock; p81br Sergey Kohl/Shutterstock; p88c Jassada watt_/Shutterstock; p89t Andrey Eremin/Shutterstock; p89c AlexLMX/Shutterstock; p89b Elena Sapronova/Shutterstock; p93c Africa Studio/Shutterstock; p112b ANNA ZASIMOVA/Shutterstock.